FROM
BABYLON TO BETHLEHEM

THE
BETHLEHEM STAR
REVEALED

By A.C. KATZ

From Babylon to Bethlehem
The Bethlehem Star Revealed

Copyright 2025 by AC Katz
Shepherds Watch Publishing

All rights reserved. No part of this book
may be reproduced, distributed or transmitted
in any form or by any electronic or mechanical methods
including photocopying, recording, or information
storage and retrieval without permission in writing
from the author or publisher.

ISBN: 979-8-218-85573-4

Printed in the United States of America

Cover design by Shepherds Watch Publishing
Photos private or public domain.

Unless otherwise noted, Scripture quotations are taken from the New King James Version. Copyright 1982 by Thomas Nelson Inc. Used by permission. All rights reserved.

King James Version
Scripture quotations marked "KJV" are take from the Holy Bible King James Version (public domain)

Scripture quotations are taken from NASB (New American Standard Bible), copyright 1960, 1962, 1963, 1968, 1971, 1972, 1973, 1977, and 1995 by The Lockman Foundation.
Used by permission. All rights reserved.

Scripture quotations marked "TLV" are taken from the Tree of Live Version. Copyright 2014 by Messianic Jewish Family Bible Society.

Scripture quotations marked "BSB" are taken from the Holy Bible, Berean Study Bible (BSB). Copyright 2016, 2020, by Bible Hub. Used by permission. All rights reserved world wide. //bereanbible.com (Berean Standard Bible).

Scripture quotations marked "NIV" are taken from The Holy Bible, New International Versuib, NIV, copyright 1973, 1978, 1984, 2011by Biblica, Inc. Used by permission. All rights reserved.

All photographs and artwork are public domain or from personal sources.

Astronomical simulations are from Starry Night software program.

Acknowledgments

Thanks to **Mondo Gonzales**, whose extensive knowledge as a Bible teacher, amateur astronomer, and astrophotographer encouraged me in my research.

A **very special thanks** to my husband, an engineer and pilot, whose years of experience in aerospace engineering, orbital mechanics, physics, and astronomy have been a constant source of support, insight, and encouragement through the many long hours of research.

Contents

Preface		i
Introduction		iii
The Bethlehem Star Supernatural or Natural?		v
Chapter 1	Apophis	1
Chapter 2	Eclipse of April 8th 2024	3
Chapter 3	Is it a Sin to watch the Heavens for a Sign?	5
Chapter 4	Scripture First	11
Chapter 5	Astronomy and Astrology vs Astrolatry	19
Chapter 6	Eclipses	25
Chapter 7	Knowledge Will Increase	29
Chapter 8	King Herod	33
Chapter 9	Caesar Augustus and the Census	45
Chapter 10	Quirinius	57
Chapter 11	Wise Men Follow His Star	65
Chapter 12	Magi	69
Chapter 13	Tradition	83
Chapter 14	Evidence	85
Chapter 15	His Star	87

Chapter 16	The Candidate	91
Chapter 17	Zodiacal Light	103
Chapter 18	The Helper	109
Chapter 19	Bethlehem	117
Chapter 20	The Star and Visit of the Magi	133
Chapter 21	What Time of Year was Jesus Born The Shepherds	153
Chapter 22	Planetary Conjunctions	159
Chapter 23	Babylonian Records Solid Evidence	167
Chapter 24	Historical Records and Dates	171
Chapter 25	The Birth of John the Baptist	191
Chapter 26	Tracing the Birth-to-Crucifixion Timeline	199
Chapter 27	Common Misconceptions Venus: Her Star- Not "His Star" The Real Bright and Morning Star	231
Conclusion: "His Star"		253
Appendix		255
Notes		267
About the Author		277

Look Up

Lift up your eyes on high: Who created all these?
He leads forth the starry host by number;
He calls each one by name.
Because of His great power and mighty strength,
not one of them is missing.
Isaiah 40:26

Can you bring forth the constellations in their seasons
or lead out the Bear and her cubs?
Do you know the laws of the heavens?
Can you set their dominion over the earth?
Job 38:32-33

It is I who made the earth and created man upon it.
It was My hands that stretched out the heavens,
and I ordained all their host.
Isaiah 45:12

Preface

Why write a book about the Star of Bethlehem? There are already many books on the subject. Most are based on tradition and writings embellished with imaginative elements, produced years later by people who were not eyewitnesses. Very few combine Scripture with astronomy, archaeology and historical writings. Some are written solely from a biblical perspective, avoiding science or history altogether, following some invisible rule that secular data is off-limits. Others rely exclusively on historical and secular sources only, leaving out the Bible. My goal in this book is to combine all available sources, with Scripture as the foundation, to determine what the Magi saw when they declared, *"We have seen His star."*

The latest discovery regarding the Shroud of Turin is causing quite a stir, spreading widely on YouTube and across the Internet. Some Christians have responded, "I don't need the Shroud, because I have my faith and I believe without it." That is perfectly fine. However, for many others, the Shroud has played a pivotal role, they have come to know and believe in Jesus through it.

People have also become believers through biblical prophecy, having learned about prophecy in the Bible, seeing how Jesus fulfilled it, and recognizing how it is presently being fulfilled all around us today. How fantastic it is, if people, seeing the new discoveries about the Shroud, turn to belief in Jesus! Many people have also become believers and realized who Jesus is through archaeology, biology, physiology, and yes, even astrology. *"What... astrology? No way, I just can't accept that. It's pagan! I am a Christian!"* Before you turn and run, let's take a closer look at the Magi.

The Magi were men of science, and they were astronomers and astrologers. While most Christians consider that bad and incompatible with faith, it's important to note it was their study of the stars that ultimately led them to Jesus, and that is good! How do Christians reconcile this? By saying the Star of Bethlehem was "supernatural", that fixes it and now we feel comfortable. But what if the star wasn't supernatural, but rather a celestial event, with an appointed time, programmed by God into the heavens during creation, which He called good?

I am not condoning astrology as it was practiced in the past, or as it is known today. Nevertheless, it is important that we acknowledge the physical characteristics and symbolism of the stars, wandering stars (Greek: *planetae),* and constellations to help us understand how they were viewed and understood by ancient observers.

Following "His Star" was not a political move on the part of the Magi, as some have suggested. The Magi made it clear that they came to see the "King of the Jews," recognizing Him as God. In their own words, they came to "worship Him." Who would undertake a long, dangerous journey from a distant land, risking their lives bearing costly gifts, simply to worship a newborn called the King of the Jews? It is likely that the Magi were influenced by Daniel and may have been descendants of the Jews who had been taken into captivity in Babylon.

The Magi had it right and correctly identified the one born as the true King of the Jews.

How did the Magi know?

To understand, we must travel back through time and examine the celestial phenomena of the ancient skies. Come along and witness the heavenly sights of long ago as we join the Magi, guided by divine signs written in the heavens, on their journey *From Babylon to Bethlehem.*

Introduction

My goal in this book is to uncover tangible evidence for the Bethlehem Star, using Scripture as the foundation combined with astronomy, archaeology and history, including writings from both biblical and secular sources. Utilizing present-day technology to determine if the Star of Bethlehem was a natural event created within the heavenly clock in Genesis 1:14, proving its trajectory was preprogrammed for its appointed time, the time of Messiah's birth.

It is possible the Bethlehem Star was a predictable event, one that occurred within normal bounds of celestial and orbital mechanics. When God uses His creation, it isn't always through unexplainable, supernatural means. My hope is to encourage people to recognize and not ignore the signs in the heavens, placed there by God for His purpose, to communicate with us. We must not be discouraged by others' interpretations or preconceived ideas about recent or previous sensationalized events, but continue to watch as the Lord has told us to do. Whether you are a pre-tribber, mid-tribber, pre-wrath or post-tribber, we are still told to watch all the signs mentioned in Scripture, which specifically includes the heavens.

In this book, you will find a compilation of information to confirm a timeline using Scripture as a foundation, along with archaeology and historical writings, tested against astronomical evidence. While there are many videos and theories about the Star of Bethlehem, some are historically, astronomically and even scripturally inaccurate. This is understandable, as many past interpretations were made without the benefit of archaeological discoveries, that are only now becoming more accessible. Over the years, due to the lack of available evidence, some have concluded, *"Well, we just can't know."* But as time goes on, archaeology and astronomy continue to shed new light and consistently confirms the accuracy of Scripture. This is just one of the many reasons we must keep watching, and not give up.

When it comes to determining the date of the Star of Bethlehem, I once read a writer say, *"It all depends on when you think Jesus was born."* I must respectfully disagree. We must be careful not to approach this subject as many have, with the presumption that we already know the date of Jesus' birth and use that as a starting point, especially when the date is based solely on tradition alone. I do consider tradition, but I do not view it as scientific evidence. While tradition has sometimes been beneficial, it has also, at times, proven to have been misleading. Still most traditions usually begin with a seed of truth. And what about the age-old traditions surrounding the Star of Bethlehem, are they true? As you read through these pages and journey along with the Magi, you will discover evidence that God planned it. It was not a myth. *"His star"* was real, and what the Bible says is true. Are you ready?

This book presents evidence while adhering to the historic, essential doctrines of Christianity: teachings that support the deity of Jesus, the triune nature of God, the inerrancy of the Scripture, and salvation by grace through faith in Christ Jesus.

The Bethlehem Star: Supernatural or Natural?

Was the Bethlehem Star supernatural or natural? Some Bible teachers believe that the Bethlehem Star was a supernatural event, something that cannot be explained by natural celestial mechanics created at the same time as the Sun, the Moon and planets (also called "wandering stars") in the beginning, as described in Genesis 1:14.

In the book *The Messenger* by Dr Thomas Horn, he investigates the possibility that the coming visitation of the asteroid Apophis in 2029 could be biblical "Wormwood" mentioned in the book of Revelation. If Apophis is indeed "Wormwood", what message does it hold? Could it be a supernatural sign in the sky?

The trajectory of Apophis follows normal orbital mechanics; it has a predictable pattern, with dates that can be calculated to forecast its next flyby. Since it is considered a naturally occurring event, should it be ignored—much like some religious teachers suggested regarding the 2024 eclipse? Or should it be considered a sign of judgment? Dr Thomas Horn was a brilliant and gifted Bible scholar who deeply understood Scripture and wasn't afraid to step outside of the man-made traditional bookends that often limit us. Did Dr Thomas Horn believe that Apophis will bring judgment? Dr Horn was clear that Apophis itself is not to be feared, but there is a coming judgment from God, and it may be connected to this celestial event. Dr Horn was not worshiping Apophis, he did not suggest that Apophis was an entity that had within itself powers or *the* power to bring judgment, yet God could use it as an instrument to bring judgment and fulfill His purpose.

Chapter 1

Apophis

Apophis has a trajectory; it follows the normal laws of celestial and orbital mechanics and moves along a predetermined, predictable path. Yet, according to some teachers, we are not even supposed to consider anything in the heavens if it fits within what they call "normal" celestial mechanics, because, in their view, God only works in supernatural ways. God preplanned certain events in the sky, the very days and years, seasons, and signs, during His creation of the heavens and the Earth (Genesis 1:14). Nature, what God breathed into existence, has become so routine to us that we have taken His creation for granted. Ask yourself, "when was the last time I got up in the morning and thanked God for the sunrise?" Some people may be schooled in Biblical Theology, and they deserve our respect, but they may have little or no background in science, math, physics or astronomy. As a result, they may not fully understand the intricacies of our universe or the delicate balance required for a solar system to exist and support a habitable planet capable of sustaining human life. Unfortunately, with the misinformation and fake science perpetuated in the recent years during the pandemic, many people have become paranoid and flee from the very word "science".

Astrophysicist Neil deGrasse Tyson said Apophis "will hit the Earth", in a 2008 interview and his 2020 book *Cosmic Queries,* about orbital mechanics and the concept of a "gravitational keyhole." The 2029 flyby will be incredibly close. Tyson's argument was that if Apophis passes through a precise, tiny point in space (a "keyhole") during that flyby, Earth's gravity could alter its orbit just enough to set it on a collision course for a future pass, such as those originally projected in 2036 or 2068. Saying "it will hit" was a dramatic way of emphasizing that the probability is not zero, and that the orbital mechanics make Apophis a serious object to watch.

The New Data: The game-changer came with the March 2021 flyby. Astronomers used the opportunity to take extremely precise radar measurements. As a result, they calculated Apophis's orbit with such accuracy that they now confidently claim it will not pass through any of the dangerous gravitational keyholes in 2029 — or so they say.

Watching the signs in the heavens is not sinning. Some teachers have suggested after the 2024 eclipse that there were some who were "ignoring the Scriptures and breaking the rules." In support of their caution about "breaking the rules," we cannot ignore Scripture and set a date for the Rapture or Second Coming when we are clearly told in the Bible "that day or hour no one knows" at least not presently. However, we are also ignoring the Scriptures if we fail to watch, as we are told to do, and yes, that includes the heavenly signs. Pastors and teachers must exercise caution and continually self-examine, ensuring they have not created their own Talmud or Catechism. Whose rules are they following? Over the years, some rules were devised by men considered extremely wise, yet they were wrong. Not only did they create their own rules, but they also distorted God's commandments and expected others to follow them. Those who did not were condemned. Many of these rule-makers and enforcers were priests or Pharisees. Since that time, some of the same Pharisaical tendencies have reemerged within the church.

What about the weather? Did God create the drought that caused the famine in Joseph's time so that, when Joseph came to power in Egypt, he would eventually bring his family to him and Israel would multiply in the land of Egypt? This was a natural event, brought on by weather patterns that caused a drought, resulting in a famine—seemingly not supernatural, but rather preprogrammed. It was part of the Earth's design, its elliptical orbit around the Sun, preprogrammed and planned from the beginning.

Chapter 2

Eclipse of April 8, 2024

After the total eclipse of April 8, 2024, there were comments circulating describing the eclipse, "X Marks the spot", as a "nothing" event, and that nothing occurred on April 8th. Compared to the wild claims some people were making, that certain events, such as the Rapture, would absolutely take place on that day, that response is understandable and was expected. Unfortunately, the claims were technically and scripturally incorrect and therefore became just another reason to ignore the format God said He created for signs (Genesis 1:14). What a horrible shame! If people fully understood the complexity involved in order for a total solar eclipse to take place, and be viewed on land in the pattern in which it occurred, they would come to realize that something most certainly did happen on April 8th, 2024.

As I mentioned in my previous book The Great American Writing on the Wall *"While eclipses are not that uncommon, total solar eclipses often occur over the ocean rather than on land. Two total solar eclipses spaced seven years apart, crossing the entirety of the United States that form an X over the largest and main Gentile nation on the earth should not be ignored -it should at least be thought provoking."*

There were those who presented erroneous information, adding a heavy dose of sensationalism, forcing Scripture into events or scenarios, claiming April 8th as the day of the Rapture or even the Lord's return to Earth. When the date came and went and whatever they claimed didn't occur, it is understandable people responded with "nothing happened." On the other hand, I find the statement that the eclipse of 2017 and 2024 was *"nothing,"* simply because they were part of "normal" creation, to be unacceptable. To suggest that *all* astronomical events or signs in the Bible falling within normal orbital dynamics, are not and should not be considered as signs, is to overlook the intentional design and purpose of God's creation. Not all astronomical events in the Bible are necessarily supernatural or beyond our ability to understand. On the following pages, I endeavor to explain my view, as well as the views of respected scholars.

Supernatural:
(of a manifestation or event) attributed to some force beyond scientific understanding or the laws of nature:
The force behind these events are attributed to God, but as time progresses we see that some are not beyond the laws of nature.

Chapter 3

Is It Sin to Watch the Heavens For a Sign?

We must be careful to not convey to people that they would be sinning just by watching the heavens, as a result they have become fearful about looking up, fearful to watch. Prophecy teachers keep watch for the latest advances in AI technology and UFO'S with no problem, I have to agree, we should be paying attention to these latest developments, they are signs and an indication of where we are on the timeline of end time events. Unfortunately, some people have become fearful, believing that learning astronomy or even predicting (oh no, that "bad" word), is a sin. Some go so far as to accuse those Christians who do study astronomy and science of being evil sinners. Yet meteorologists (the weathermen) predict the weather, and the Moon and the Sun do, in fact, affect Earth's atmosphere and weather patterns. The National Oceanic and Atmospheric Administration or NOAA, "Every day, the NOAA's supercomputers collect and organize billions of earth observations, such as temperature, air pressure, moisture, wind speed and water levels, which are critical to initialize all numerical weather prediction models. "The calculations they make are the foundation of NOAA's life-saving weather predictions". Are Meteorologists sinners for predicting the weather?

Luke 12:55-56 *And when the south wind blows, you say, 'it's going to be hot', and it is. Hypocrites! You know how to interpret the appearance of the earth and the sky. How is it that you don't know how to interpret this present time?"*

You are not sinning by watching, studying, or predicting the path of the Sun, Moon, stars, or planets based on their patterns, in fact, we are told to do so in the Bible. However, the Bible makes it clear, that it **would** be sinning to personifying and deify the

Sun, Moon and the stars, treating them as gods, (deifying stars, balls of Hydrogen gas, planets or as the ancients called them, wandering stars, comets, meteors etc.), believing they are responsible for bringing judgment upon the Earth. That they are sentient beings capable of thought and can be offended, or that they possess power and control over our lives.

The study of astronomy is not a sin; it is a science. As for astro**logy,** it depends on the definition, of which there are several, because astro**logy** has been perverted in many ways. If someone believes that the heavens, Sun, Moon and stars have power, or that they are gods and they decide themselves, to bring about judgment, resulting in them fearing and worshiping these celestial bodies, then yes, according to Scripture, this would be considered sin. Their fear is based on their personification and idolization of the heavens, as referenced in the following Scriptures.

Jeremiah 8:2 *They will spread them out to the sun, the moon and all the host of heaven, which they have loved and served, and after which they have walked and sought, and which they have* ***worshiped.***

Jeremiah 10:5 *"their idols cannot speak. They must be carried because they cannot walk! Do not fear them for they can do no harm nor do any good."*

1 Chronicles 16:25 *For great is Adonai and greatly to be praised. He is to be feared above all gods. For all the gods of the peoples are idols,* ***but Adonai made the heavens.***

Chronicles 33:2-5 *(BSB)* Speaking of King Manasseh's transgressions, *"He did evil in the sight of Adonai, just like the abominations of the nations that Adonai had driven out before Bnei-Yisrael. For he rebuilt the high places that his father Hezekiah had demolished, he re-erected altars for Baalim, made Asherah poles,* ***and bowed down to all the host of heaven and worshiped them.*** *He built altars in the House of Adonai of which Adonai had said, "My Name will be in Jerusalem forever."* ***He also built altars for all the host of heaven in the two courtyards of the House of Adonai.***

Studying science and allowing it to predict or making a prediction as a result of that study, is not a sin. We can predict the weather; we know when it is going to be spring due to the elliptical orbit of Earth around the Sun.

Consider the Medical field: we look at synovial fluid within joints and know when it is low, it can lead to cartilage damage, osteoarthritis, and joint pain. The loss of synovial fluid results in increased friction within the joints, leading to their degeneration. When the fluid loses its viscosity, it can cause deterioration. The use of Hyaluronic Acid injections can be helpful to increase the viscosity of the fluid in the joint and can help with lubrication, decrease the pain and increase the longevity of the joint and tissues.

Doctors use CT to show bone density or the lack thereof. By studying those results, they can predict that a person who has the bone density of pumice or Swiss cheese, will likely suffer fractures within a predicted time frame, possibly within months or a few years.

Why is predicting through one science acceptable, yet predicting something through astronomy not acceptable? We can predict with almost certainty the path of totality for an eclipse and the trajectory/pathway of an asteroid or comet. But I want to make clear: predicting a trajectory is one thing; predicting its meaning or the message God has for us in it, is another thing.

Some teachers are claiming we can only look for signs in the heavens during the Tribulation, specifically on the very "Day of the Lord" asserting that the Day of the Lord is limited to only a 24-hour period. Although the reference to the Sun and Moon being darkened in Matthew, applies to a time during the Tribulation, it is a time lasting longer than a 24 hours. It clearly says in Matthew 24:29 *"But immediately after the trouble of those days, the sun will be darkened and the moon will not give its light and the stars will fall from heaven and the powers of the heavens will be shaken."* After the trouble of what days? After all the troubles listed in Matthew 24:3-15.

In Acts 2:1-20 we read signs in the heavens are not limited to the Tribulation period alone. When Peter quotes the prophet Joel, Acts 2:16 *This is what was spoken about through the prophet Joel: and it shall be in the last days, says God that I will pour out my Spirit in all flesh. Your sons and daughers will prophesy"*. This was taking place that very morning. He goes on to say *"And I will give wonders in the sky above and signs on the earth beneath"* to be specific it says, *"<u>before</u> the coming great and glorious day of the Lord."* If it it says it will be *before* "the Day of the Lord" then it's *before,* not *on* the Day of the Lord. However, there are Scriptures that explain "the Day of the Lord" as a time period that is not limited to a 24-hour day. In the above Scriptures, Peter was referring to the time beginning from that moment—Shavuot/Pentecost forward—when the Church received the Holy Spirit. It is a clear explanation by Peter that they were already in that time, the time when the Holy Spirit was given, and it was evident, by the speaking in tongues taking place at that moment.

Although we are told, there will be signs in the heavens, we must be careful not to be distracted by what can be confirmed as mere sensationalism. Instead, we should seek to understand what truly qualifies as a rare astronomical event, one that clearly falls within the category of a biblical sign.

If looking at the heavens and viewing scientific data is sinful and breaking the rules, then we have to ask, "What about the Magi?". If, according to rules set by some Bible teachers that "they broke the rules", then how do those teachers reconcile this without accusing the Magi of being rule-breaking sinners? Some go even further, accusing them of more than being just breaking rules, without any evidence or applicable Scripture regarding the specific Magi mentioned in the Gospel of Matthew. To pacify their confusion and support their conjecture, they justify the Magi's actions of looking up and following the star "His star" by claiming the "Star of Bethlehem was a supernatural event".

But what if it was not a supernatural event? And if we agree with the methods of the Magi, or even acknowledge the science of astrology, are we sinners too?

Although there were many interpretations of Magi historically, some Magi in the Mediterranean area had established a sound reputation for both character and learning. Matthew indicates no value judgment as to the characters of the Magi that visited Jesus.

While some teachers claim that the Star of Bethlehem could only have been a supernatural event, they cannot confirm that it truly was. In fact, there is far more archaeological and astronomical evidence indicating that it occurred within the planned and predictable laws of celestial mechanics than evidence supporting a purely supernatural explanation. Unfortunately, most of what people believe about the Star of Bethlehem has come from Christmas cards, carols, and nativity scenes, ideas that have been planted in our minds since childhood. Yet there is more historical evidence for the existence of Jesus than there is for Alexander the Great, but everyone accepts that Alexander existed without question. **There is more proof that the Star of Bethlehem was a natural event,** rare but still natural, scripturally, astronomically, archaeologically and historically, **than there is proof it was supernatura**l!

The Star of Bethlehem was a planned event, with an appointed time, designed by God within the natural celestial mechanics He established at creation. The Magi were exceptionally wise men, the very best scientists, astronomers, astrologers of their time, as they had been for hundreds, even thousands of years. The Magi said in Matthew "we saw His star in the east" in contrast to we saw <u>a</u> star, this is a clear indication that the Magi knew there was one particular star associated with the "King of the Jews". They did not sin by watching and predicting what it meant. They did not sin, they were not fearful, they did not personify the planets, they knew the stars and planets were not actual beings that decided our fate. Yet they knew it was a sign; they rejoiced and were motivated to take action.

*The heavens declare the glory of God;
the skies proclaim the work of His hands.
Day after day they pour forth speech;
night after night they reveal knowledge.*

Psalm 19:1-2

Chapter 4

Scripture First

For what purpose did God say He created the Heavens?
The heavens are God's great clock—a divine timepiece with gears and hands across its face. It is set above us where all can see it, and when properly understood, those who study the clock can discern where we stand on God's timeline.

Genesis 1:14 And God said, "Let there be lights in the expanse of the sky to separate the day from the night; and let them be for signs (Le Ot), and for seasons (moedim), and for days and for years."

Mondo Gonzales the Director of the **Psalm 19 Project** and co-host at **Prophecy Watchers**, in the June 2024 issue of The Prophecy Watcher magazine takes a close look at Genesis 1:14 and explains the five purposes for the heavens. #1 to separate the day from the night, # 2 let them be for signs, #3 and for seasons, #4 and for days #5 and years. "We know from the summary of verse 16 that these lights were the Sun, Moon and the stars. We must remember that the 5 purposes were all given in the context before evil was in the world. He created all of these and called them good (1:18)." Mondo goes on to explain, "The best lexicon used by biblical scholars is *The Hebrew and Aramaic Lexicon of the Old Testament,* and it lists 9 different nuances of the Hebrew word for sign. It isn't relevant to list all nine but I will reference two of the nuances because they apply to our text in Genesis 1:14. In the context of that which is good and not evil, one of the main nuances for this word *sign* in Hebrew is that it is used to refer to an object lesson as a remembrance or memorial for God's goodness and kindness." "When God says that the starry skies would be used of a sign/memorial, it makes sense that from the beginning of the creation the constellations or what God calls the

Mazzaroth (Zodiac) would be seen as a memorial to His creative power (Job 38:32)."

Definition: "signs" (the Hebrew word *Ot)* a signal (literally or figuratively) such as a flag, beacon, monument, prodigy, evidence, etc. - mark, miracle, sign, promised by prophets as pledges of certain predicted events. The word *sign* is a noun; a clue that something happened, or a display that communicates a message. A sign is an object or event whose presence or occurrence indicates the probable presence or occurrence of something else. A sign suggests the presence or existence of a fact, condition, or quality, or gesture used to convey an idea, a desire, information, or a command. A sign is a symbol or message in a public place that gives information or instructions.

Hebrew **Moedim** appointed time(s) Moed "for seasons" specifically, a festival; conventionally a year; by implication, an assembly (as convened for a definite purpose) the place of meeting; also, a signal appointed time, place of solemn assembly, set solemn feast, appointed due season.

   The beginning of the book of Daniel through Daniel 2:3 was originally written in Hebrew. Daniel 2:4 to 7 was written in Aramaic the language of Babylonians. Chapters 8-12 are written in Hebrew.
    In the book of Daniel Chapter 11:27 it says, "for the end will still come <u>at the appointed time</u>", the words "appointed time" are used and were originally written in Hebrew as "la moed".

Daniel 11:29 *"<u>At the appointed time</u> (moed) he will invade the south again, only this time will not be as before.*
Daniel 11:35 *Even some of the wise will stumble, so that they may be refined, purified and made spotless until the time of the end, for it will still come <u>at the appointed time</u>* (la moed).

   In the above Scriptures the "appointed time" was originally written in Hebrew as "la moed" the mention of "moed" above is not used in reference to a particular Jewish feast day such as Passover, Shavuot (Pentecost) or Yom Kippur etc. Although the word *moed* is often translated as "feast" or "festival," this is not always entirely accurate. The term "appointed time" does not necessarily mean that Scripture is referring to one of the seven feasts of Israel, as is evident in the passages cited above. *Moedim* = feast, festival, meeting time, appointed place, appointed time, time, times, appointment, a time God decides or has planned on His calendar and is not restricted to a feast, holy day or festival.

Moed has more than just one meaning, therefore, to assume that the word "moed" in Genesis only refers to feast days, is not a complete and thorough interpretation of the Scripture. *Le Ot* = signs, are to let us know when certain events will take place, whether they are feast days, or other moeds/ appointed times spoken of in the above Scriptures in Daniel.

Habakkuk 2:2-3 *Then the LORD replied: "Write down the revelation and make it plain on tablets so that a herald may run with it. For the revelation awaits an appointed time; (la moed) it speaks of the end and will not prove false. Though it linger, wait for it; it will certainly come and will not delay.* (NIV)
Jeremiah 8:7 *Even the stork in the sky knows her appointed times (moed), and the turtledove, swallow and crane observe the time of their migration, but My people do not know the judgments of Adonai.* (TLV)

The above Scriptures are additional examples of the word *moed* (appointed times) being used when it is not referring to Jewish holy days or feast days.

In 2 Kings 20:8-11, Isaiah 18:6-8 these events can presently be understood as supernatural and not within natural orbital mechanics. They are however, astronomical signs (Ot) not limited to feast days (moeds).

The heavens were created by God to be viewed by us and evoke awe-inspiring wonder about their Creator. However, they were created for more than one purpose, as is seen throughout the Bible. Not only do they reveal the glory of God, but they are a constant witness to us that God is in control of creation and events, past, present and future, natural or supernatural.

Psalm 19:1-5 *The heavens declare the glory of God; the skies proclaim the work of His hands. Day after day the pour forth speech; night after night they reveal knowledge. There is no speech or language where their voice is not heard. Their voice has gone out into all the earth, and their words to the end of the world.*

The above Scripture lends additional meaning to the purpose for which the heavens were created. They speak and reveal knowledge and when they are properly understood, they tell us where we are on the biblical timeline, whether it's a feast day, or the supernatural sign in the sky as in 2 Kings 20:8-11, Isaiah 18:6-8, and the Bethlehem Star Matthew 2:10.

Psalm 19:1-5 proves that the speech of the heavens and message is for everyone on the Earth, including Gentiles. It is not only for viewing the sky to schedule and observe Jewish feast days. It is also evident by the previous events in the heavens, such as the Bethlehem Star, the Revelation 12 Sign, and end times described in Matthew 24:29 and Luke 21:25-26, that God will use nature and astronomical events for signs.

A Simple Analogy But Helpful

When we look at the face of the analog clock, there are arrows or hands, in various sizes that point to numbers on the face of the clock as they move. The shorter hand is the hour hand and conveys a message, it lets the viewer know what hour of the day it is. The longer hand is called the minute hand, it lets the viewer know how many minutes after or before the hour it is. The tiny often metallic hand also has a purpose and a message, it points to the seconds ticking by. I know this sounds like a juvenile analogy, but there is a generation that is unable to tell the time on an analog clock, likewise many of us in this generation are unable to read the clock in the heavens.

As we read in Genesis 1:14 one of the purposes God created the heavens was for Jewish feast days (moedim) or appointments with God throughout the year, appointment times. God told the Israelite's in Scripture when (on His clock) these appointment times are scheduled and are to be observed. Mandated by the Bible, Passover is to be celebrated by the Jewish people as an annual commemoration of their Exodus out of Egypt. Passover is an appointment time with God; it is marked by a particular time on the clock in the heavens. It is observed during the springtime and is to be celebrated at the new moon during the Hebrew month of Nissan. The date of Passover changes each year because the day is not set by the Gregorian calendar but by lunar based orbital mechanics. The Jews know the time when it is to be celebrated by watching the clock, by observing the pattern and pathway of one of the hands on that great cosmic clock, the Moon.

The duty of the Levites in helping the descendants of Aaron in the Avodah (service of) God's House.

1 Chronicles 23:30 *"and to stand every morning to thank and praise Adonai. They also did this in the evening and whenever burnt offerings were offered to Adonai <u>on Shabbatot, New Moons and the Moedim.</u>"*

New Moons: It takes about 29.5 days to go from one new moon to the next new moon, by understanding the pattern and timing they could count out the days and they could prepare.

Psalm 104:19 *He made the moon for appointed times, the sun knows it's going down.* (TLV)

The new moon in the month of Nissan is the appointment time for Passover, which was previously scheduled in the gears of the clock by God. The other "moedim" feasts, are appointments scheduled for other times on the clock throughout the year. In order for the Jews to know when those times are, they must continue to watch the clock, just like we would to see if our doctor appointment is in 2 hours or 15 minutes.

Jesus admonished the Pharisees, they knew how to read the heavenly clock for "the seasons" for the weather and crops, but did not know how to apply it to signs and prophetic declarations or as Jesus said, "they would have known the time of His coming".

Luke 12:54 *He said to the crowd: "When you see a cloud rising in the west, immediately you say, 'It's going to rain,' and it does. And when the south wind blows, you say, 'It's going to be hot,' and it is. Hypocrites! You know how to interpret the appearance of the earth and the sky. How is it that you don't know how to interpret this present time? (NIV)*

Matthew 24:43 *But know this, that if the master of the house had known what hour the thief would come, he would have watched and not allowed his house to be broken into. (NKJV)*

Mark 13:28- 37 Jesus tells us to "watch."

Supernatural
Sometimes the astronomical signs in the heavens spoken of in Scripture were supernatural, as when the Sun stood still in Joshua 10:13-14 *On the day that the LORD gave the Amorites over to the Israelites, Joshua spoke to the LORD in the presence of Israel: "O sun, stand still over Gibeon, O moon, over the Valley of Aijalon." "So the sun stood still and the moon stopped until the nation took vengeance on its enemies. (is it not written in the book of Jashar?) So the sun stopped in the middle of the sky and delayed going down about a full day." There has been no day like it before or since, when the LORD listened to the voice of a man, because the LORD fought for Israel....*

Isaiah 38:8 *I will make the sun's shadow that falls on the stairway of Ahaz go back ten steps." So the sunlight went back the ten steps it had descended.*

In the above Scriptures, it is evident that these were both supernatural events. I believe God simply pulled the hands back on His clock, just like we do when we change our clocks for daylight saving time. (Isaiah 38:8) In the case with Ahaz, God moved the large hand (the Sun's shadow) back 10 steps on the stairway of Ahaz. These astronomical gymnastics would be considered unnatural in the timing of the orbital mechanics of the Sun.

The heavens were designed and created by the great clock maker, better than the famous Swiss watchmakers and with perfect accuracy.

"God is working a plan, because God's plans are linked to time, we need to be paying attention to what the skies are doing, because they might tell us something about what God is about to do." Dr Michael Heiser

Chapter 5

Astronomy and Astrology vs Astrolatry

If you work in the medical field, as the majority of my family has for generations, you know the importance of understanding Latin/Greek suffixes, in order to understand illnesses, conditions, medical and surgical procedures in the science of medicine. They are also helpful to understand terms in science in general and throughout our daily language.

Understanding suffixes
Suffixes are used to identify the full meaning of a particular word. Below are examples of familiar suffixes used in medical terminology.
Suffixes
itis = inflammation of
ectomy = surgical removal of
otomy = incision or cutting
ostomy = cutting a hole and leaving an outlet
plasty = repair

The word *appendix* with the added suffix "-itis" then becomes *appendicitis*, which means inflammation of the appendix. The suffix "-ectomy" added to the word *appendix* is *appendectomy,* which means the surgical removal of the appendix. A similar use is *tonsillitis* and *tonsillectomy.*

The suffix- logy - means a branch of learning, or study of a particular subject, theory, doctrine, science, "Field of study". Greek *logia*, from *log*, "to speak, tell;" The suffix *logy* as in geology means the study of the solid earth, such as rocks and minerals. That same suffix *logy* and the Greek root **word *astron* "star" becomes astrology and also means the study of** the relation of significant celestial moments or movements.

The suffix – nomy -*nomos*- means a system of rules or laws, or body of knowledge of a particular subject. Greek *astronomos,* literally "star-regulating," from *astron* "star" and *nomos* "arranging, regulating; rule, law" or "science of".

Astrology was originally identical with astronomy, including scientific observation and description. The special sense of "astronomy applied to prediction of events" was divided into natural astrology "the calculation and foretelling of natural phenomenon" and judicial astrology "the art of judging influences of stars and planets on human affairs." By late 17th century, astronomy came to mean exclusively "the scientific study of the heavenly bodies."

The suffix – latry – means "worship of" worship, service paid to the gods.
Astro**latry** meaning worship of the heavenly bodies
　　The worship of heavenly bodies is the veneration of stars, planets or other astronomical objects as deities. The stars or all the heavenly bodies, whether alone or jointly have at one time or another, been the object of worship. It is well-known, the Ancient Egyptians worshiped the Sun which they called "Ra". Many early religions included the worship of the Moon and when called *Luna* the term is usually a reference to a personified Moon.
　　We study the movements of the Sun and Moon and constellations and note how celestial movements affect our weather and it is predicted and recorded in a Farmers Almanac or a written or viewed on TV as the weather report.
Astronomy and Astrology are not technically religions; both are the study of what was created by God, and in Genesis 1:18, God calls it good!! It is no more evil than studying a pine tree or a rock. It's not a religion unless you make it one.

Genesis 1:16-18 *God made two great lights: the greater light to rule the day and the lesser light to rule the night. And He made the stars as well. God set these lights in the expanse of the sky to shine upon the earth, to preside over the day and the night, and to separate the light from the darkness. And God saw that it was good....* (BSB)

Jeremiah 10:1-13 speaks about idolatry, note the suffix "latry" meaning "worship of". In the Scriptures below we read about a type of idolatry....Astrolatry

Jeremiah 10:2 *Thus says Adonai: "Do not learn the way of the nations or be frightened by signs of the heavens though the nations are terrified by them.*

Jeremiah 8:2 *They will spread them out to the sun, the moon and all the host of heaven, which they have loved and served, and after which they have walked and sought, and which they have* **worshiped.**

Deuteronomy 4:19 *And lest thou lift up thine eyes unto heaven, and when thou seest the sun, and the moon, and the stars, even all the host of heaven, shouldest be driven to* **worship them, and serve them,** *which the LORD thy God hath divided unto all nations under the whole heaven.*

    Looking for signs in the sky is not a sin, nor were the Magi great sinners for having done so and recognizing the meaning of "His star". The sin of the nations was that they worshiped and feared the heavens, dreading them because they saw the Sun, the Moon and stars as gods who could think and had the power to bring about judgment and punishment. They had personified the heavenly bodies and lived to appease them through worship and rituals. They worshiped the created instead of the Creator; they feared the heavens and did not fear God.

Personify: *represent (a quality or concept) by a figure in human form: "public pageants and ceremonies in which virtues and vices were personified" attribute a personal nature or human characteristics to something nonhuman: represent or embody in a physical form.*

    The other "nations" personified the heavenly bodies and

created images of them in human form. Babylonians, Egyptians, Romans, Greeks, and others created numerous statues and placed them in temples dedicated to the worship of their false gods.

In Jeremiah 10:5 it says *"their idols cannot speak. They must be carried because they cannot walk! Do not fear them for they can do no harm nor do any good."*
It goes on to say in Jeremiah 10:7 *There is none like You, Adonai! You are great and great is Your Name in power. Who should not fear You, Ruler of the nations? For it is your due! For among all the wise of the nations and in all their kingdoms there is none like You.*

Jeremiah 10:11-12 *Thus you will say to them: "The gods which did not make the heavens and the earth will perish from the earth and from under the heavens." He made the earth by His power, established the world by His wisdom and stretched out heaven by His understanding.*

Revelation 14:7
And he said with a loud voice, "Fear God and give him glory, because the hour of his judgment has come, and worship him who made heaven and earth, the sea and the springs of water."

An example of Astro**latry** is in Chronicles 33:2-5 Speaking of King Manasseh's transgressions, *"He did evil in the sight of Adonai, just like the abominations of the nations that Adonai had driven out before Bnei-Yisrael. For he rebuilt the high places that his father Hezekiah had demolished, he re-erected altars for Baalim, made Asherah poles, and* **bowed down to all the host of heaven and worshiped them.** *He built altars in the House of Adonai of which Adonai had said, "My Name will be in Jerusalem forever."* **He also built altars for all the host of heaven in the two courtyards of the House of Adonai.** *(BSB)*

Astrology (the study of) in ancient times and throughout history has been perverted, as Satan so often does with all that God calls good (Genesis 1:18). It is always the things that carry such importance to God, and that which God calls good, that the Devil tries to pervert, tricking the minds of Christians into believing erroneous teachings, causing them to ignore the very things God uses to bring Him glory and communicate with us, such as the heavens, God's great timepiece! The devil does not want you to watch or understand the signs in the sky, the very format God said He created for "le Ot" (signs) and seasons, and for "moed" (appointed times— not only feasts), as well as days and years. Don't stop looking up at God's time piece—the one He designed, engineered and calls good. Continue to worship the Creator, not the created, and give Him the glory!

Isaiah 40:26 *Lift up your eyes on high: Who created all these? He leads forth the starry host by number; He calls each one by name. Because of His great power and mighty strength, not one of them is missing.*

We Are Commanded to Watch.

God has scheduled certain events to occur at particular times on the astronomical clock, just as He has scheduled moedim (feasts) at other times on the clock. We don't always know what God has scheduled for a certain day, month or year, but we are told to *WATCH*.

God does tell us about some of the events scheduled in His appointment book— the Bible. They were already scheduled when He created the heavens. Sometimes, we are told exactly when something will take place, as in the Book of Daniel, when the Messiah would enter Jerusalem, which they could calculate from the decree of Artaxerxes to rebuild Jerusalem. If they had been watching the heavenly clock and properly counted the days, months and years, they would have known the day of His arrival, the day Jesus entered Jerusalem, on what is known as Palm Sunday. But the religious leaders missed it, they were too wrapped up in their own arguments, concerned with their own rank in a hierarchy of prestige, stubborn in their traditions and they were not watching.

Daniel 9:25 *"So know and understand: From the issuing of the decree to restore and to build Jerusalem until the time of Mashiach, the Prince, there shall be* ***seven weeks and 62 weeks****. It will be rebuilt, with plaza and moat, but it will be in the times of distress. Then **after the 62 weeks Mashiach will be cut off** and have nothing." (TLV)*

In the Book of Daniel and other Scriptures, we are encouraged to count days, weeks and years, and understand other events.

Daniel 12:11 *"From the time that the daily sacrifice is abolished and the abomination that causes desolation is set up, there will be 1,290 days. Blessed is the one who waits for and reaches the end of the 1,335 days."*

Revelation 12:3 says, *"And I will appoint my two witnesses, and they will prophesy for 1,260 days, clothed in sackcloth."*

There are times throughout Scripture we are told about an event that is to come in the future, but we are not told at what time it will occur. God has many appointments scheduled, but we don't always understand their full meaning, at times only the prophets were able to see clearly. Watching the heavens (the heavenly clock) is not a sin, trying to figure out the meaning is also not a sin, attempting to understand the timing of events is also not a sin. We are told to *watch*, and as we are led by Scripture, attempting to understand the timing by observing creation is being obedient. God never told us to ignore the heavens, He called it good. When we wear a wristwatch, we "watch" it. We don't believe it has the power to bring judgment, we don't worship it or believe it is a living being, but we watch it in order to know what time it is.

Chapter 6

Eclipses

The Sun, the Moon and planets are hands on the astronomical clock, during certain times they line up and an eclipse occurs and can mark an event in history.

We know historically that King Herod died in between a lunar eclipse and Passover, as it was recorded by the Roman Historian Josephus (*Antiquities of the Jews*). There are only two times that a lunar eclipse occurred toward the end of Herod's reign as king, 4 BC and 1 BC. It marked the time when Herod died, we are given the marker, the eclipse and we can read the astronomical clock to tell us the time when it occurred. Pagans take it a step further but in the wrong direction, they believe the eclipse caused or brought about the death of the king.

An eclipse is within the natural orbital mechanics specifically created by God for our viewing from Earth. Does an eclipse bring judgment? It doesn't bring judgment, it doesn't decide, but it is another hand on the clock in the heavens, similar to the hands of an analog clock that points to 4:00 two times per day, 4 p.m. and 4 a.m. Eclipses occur on the average twice per year, sometimes more and in various ways such as Solar eclipses, Lunar, Partial, Annular Hybrid.

Just as the Sun, the Moon and stars tell us it's springtime and time to plant, or the end of summertime to harvest, the "days, months and years" tell us what the weather will likely be, hot, cold, rainy etc.

An eclipse is also designed into the normal celestial mechanics. It is remarkable that the Sun is 400 times larger than the Moon and 400 times farther away, making it possible for partial and total eclipses to be viewed from Earth. The number 400, in the Hebrew alphabet is represented by the letter *Tav*. The paths of the 2017 and 2024 eclipses made a mark across the United States, that mark was an "X" a *Tav*.

Ezekiel 9:4 *The Lord said to him, "Go through the midst of the city, throughout all of Jerusalem, and put a mark on the foreheads of the men who sigh [in distress] and grieve over all the repulsive acts which are being committed in it."*
For the slaughter of the idolaters was about to begin.

The Tav

The Tav and its meaning are confirmed by Scripture. The Tav was evidenced by the path of totality for the April 8, 2024 eclipse crossing the path of totality of the eclipse of the August 21, 2017 which had occurred only seven years earlier. Both of their trajectories and paths of totality made a Tav, a mark across the United States. Does that mean the eclipse brings judgment? No, but it certainly is a marker, marking a location, and a point in time.

Biblical TAV

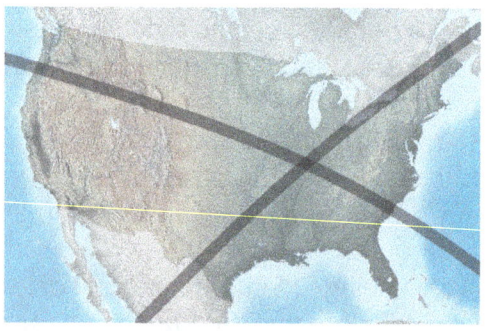

Do we know for certain what Tav means? It is the Hebrew word for the last, the end, in Greek it is Omega. Revelation 22:13 *"I am the Alpha and the Omega, the First and the Last, the Beginning and the End."*

Revelation 21:6 *"He said to me, 'It is done. I am the Alpha and the Omega, the Beginning and the End"*

Omega is the 24th, and last letter in the Greek alphabet, it also means "the final part" or "the end". The book of Revelation, is the final book of the New Testament, as well as the final book in the entire Bible.

Many people are as frightened by astronomy as they are the Book of Revelation, but it clearly gives hope to those who follow the Lord. The Book of Revelation is meant to be informative; it explains what is coming on the Earth and in the heavens, the final judgment, the return of Jesus and is a reminder of a glorious future. We are allowed, in fact told, to watch all the events taking place pointed out in Scripture, including watching Gods celestial clock. However, we are clearly told we cannot know the day or the hour of the Rapture or Second Coming.

Chapter 7

Knowledge Will Increase

In the Book of Daniel, it says knowledge will increase. For many years, there have been claims that King David never existed. Yet archaeology has now proven, by the discovery of Mesha Stele, that King David did exist. In 1868, a large stone was uncovered in the biblical city of Dibon. It recorded the victories over the Israelites by Mesha, the king of Moab. These same battles were recorded in the Bible. Line 31 on the Mesha Stele *"The **house of David inhabited Horonaim**. Line 32 and Chemosh said to me: "Go down! Attack Horonaim." So I advanced against it,"*. Numerous other lines in the Mesha Stele correlate with the Bible as it mentions more names of those in the Bible involved in the very same events.

    People didn't believe there had been an actual pool of Siloam, yet recently archaeologists have uncovered it. The Pool of Siloam mentioned in John 9:1-11 where Jesus healed the blind man, was rediscovered during construction work for the reparation of a large pipe which was located at the southern end of the City of David, in the autumn of 2004. Archaeologists Eli Shukron and Ronny Reich identified the two ancient stone steps, the 225-foot-long structure they found upon further excavation confirmed that they were part of a magnificent pool from the Second Temple period in which Jesus lived.

Additional recent Archaeological Discoveries The ivory pomegranate found at Tel Shiloh, the Beka weight from the Temple Mount and the fortification wall at Lachish. From the temple mount sifting project, five rare coins from Jerusalem bearing the inscription in ancient Hebrew (Judah). They are believed to be some of the earliest evidence of Jewish coin minting in Israel from the fourth century BC, from around the time described in Ezra and Nehmiah. This period is recorded as the time of the return of the Jewish people from Babylonian exile by decree of Cyrus the Great, to begin construction of the Second Temple. The limestone block with the Pilate inscription found in June 1961 discovered in ancient Roman ruins, the block bears the inscription "Pontius Pilate prefect of Judea" before this discovery scholars questioned the historicity of the man who ordered the death of Jesus. The only previous evidence of Pilate's existence was known through the Christian Gospels and the writings of the Jewish historian, Josephus.

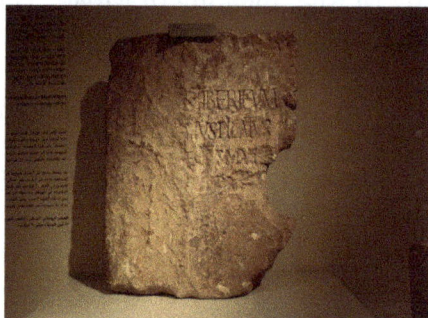

Photo Credit BR Burton

On the block is a dedication to the deified Augustus.

To the Divine Augusti [this] Tiberieum
...Pontius Pilate
...prefect of Judea
...has dedicated [this]

Pontius Pilate inscription; the original stone, now located in the Israel Museum, Jerusalem.

Photo credit Israel Antiquities Authority

The Bethlehem Bulla – the earliest archaeological evidence that Bethlehem existed during the First Temple period.

Discoveries in science, due to advances in technology, have increased exponentially. With the technology now available, we have been able to learn that events in the Bible we previously could not prove or explain, are no longer deniable. Some things remain speculation, while other scientific discoveries provide tangible evidence.

Chapter 8

King Herod

Beginning with Scripture to Establish a Timeline for Further Investigation.

Matthew 2:1-3 *Now after Jesus was born in Bethlehem of Judea, in the days of King Herod, the Magi from the east came to Jerusalem, saying, "Where is the one who has been born King of the Jews? For we saw His star in the east and have come to worship Him." When King Herod heard he was troubled and all Jerusalem with him.* (NASB)

King Herod "The Great"
Born 74 BC Died March – April 4 BC In Jericho
Ruled 40/37 BC – 4 BC
Not to be confused with his son Herod Antipas Tetrarch over Galilee (reign 4 BC – AD 39) who killed John the Baptist, (Mark 6, Luke 23:7-12, Acts 4:27).
In Acts 12: The persecutor of the apostles is Herod Agrippa l.

    Herod's religious and ethnic background was complicated. He was Nabatean from his mother's side, a Nabatean Princess from an Arabic tribe in southern Jordan, while his father was an Idumean-Jew from a family of Idumean converts. He had both a Hellenistic and Jewish upbringing. Understandably, he was the embodiment of religious identity crisis, marked by contradictions and paranoia. Preceding Herod's reign, Antigonus II was a true King of the Jews, not just High Priest. Herod did not take the throne as King of the Jews until King Antigonus died in 37 BC. However, Herod had been given the title "King of the Jews" by the Roman Senate in 40 BC. Herod was not born King of the Jews— he was appointed.

Why was this King Herod called "The Great"?

Herod the Great was the King of Judea, responsible for large-scale building projects, including remodeling the Jerusalem temple. His ambition was unlimited, driven by pride, fame and insecurity. He created masterpieces that were symbols of power and majesty, temples and fortresses that remain architectural marvels of the world to this day. Herod tried to appease his Jewish subjects and religious leaders by the renovation of the most important building to Judaism.

The Jewish Temple: The renovation of the Jewish temple in Jerusalem, the most well-known of Herod's architectural masterpieces and symbol of Herod's influence and lavishness of his reign.
Masada: The famous and formidable Fortress.
Herodium: Herod's unique and magnificent palace, designed in the shape of a mountain, to display his wealth and power. The living quarters and banquet halls were lavishly decorated with elaborate mosaic floors, ornate columns, frescoes and exquisite architectural detail.
Caesrea Maritima: A magnificent harbor and ancient port on the Mediterranean coast of present day Israel. Named after the Emperor Caesar Augustus, Caesarea had an artificial harbor of large concrete blocks typical of Hellenistic-Roman public buildings. Herod designed and constructed a world class harbor, including architectural marvels such as a grand amphitheater for entertainment and Hippodrome for chariot races.
Antonia Fortress: Strategically placed it stood next to the Temple, housing highly trained guards, it was a formidable reminder of power.
Aqueducts: A means of transporting a reliable supply of water to the city, in Caesarea Maritima, Jerusalem and Herodium.
The Jericho Winter Palace: Herod's stunning and luxurious vacation home.

The story in Matthew is placed during the time of King Herod the Great (74–4 BC), who was appointed "King of the Jews" by the Roman Senate at the urging of Mark Antony in 40 BC. (Josephus, *Jewish War* 1.14.4).

Herod's father, Antipater I, had been appointed "Steward" of the region by no less a figure than Julius Caesar himself. (Josephus, *Jewish Antiquities* 14.143). After his father died, Herod, with the blessing of the Imperial family took over. The Roman senate entrusted the throne of Judea to Herod the Great at the close of 40 BC the same year as the Parthian conquest. Josephus puts this in the year of the consulship of Calvinus and Pollio (40 BC).

Evidence vs Speculation for the Date of Herod's Death.

Herod's Death- Astronomical data in Josephus's account of the Lunar eclipse and Passover which occurred two times in 4 BC and 1 BC, which one was during the time of Herod's death?

It is recorded by the Roman historian Josephus *Antiquities* XVII:6:4 who writes that Herod died between a lunar eclipse and Passover. If we look at astronomical records from 8 BC to 1 BC there were two lunar eclipses during the latter portion of King Herod's reign. There was one lunar eclipse in 1 BC and another lunar eclipse in March 4 BC. We can look at the heavenly calendar an astronomical program called Starry Night to confirm the eclipses of 4 BC and also 1 BC.
The lunar eclipse of March 13, 4 BC viewed from Jerusalem to beginning of Passover April 11, 4 BC.
The lunar eclipse of January 10, 1 BC viewed from Jerusalem to beginning of Passover April 8, 1 BC.

The consensus among most scholars is 4 BC is the date of Herod's death based on various historical data. However, there has been some argument over the lunar eclipse of 4 BC and the lunar eclipse of 1 BC as to which lunar eclipse Josephus was referencing to mark the date of Herod's death. Those presenting this idea do so with a predetermined date for the birth of Jesus. The basis of the argument being because the lunar eclipse of 1 BC was a total lunar eclipse, assuming it would have been spectacular and very noticeable, compared to the 4 BC eclipse being a partial lunar eclipse assuming it was not as noticeable. Evidently, there is confusion on the part of those presenting this as being significant. It would make sense if these were two Solar eclipses, but these were two LUNAR eclipses.

In a solar eclipse, the lineup is the Earth then Moon and Sun, the Moon traverses in between the Earth and the Sun. A lunar eclipse occurs when the Sun, Earth and the Moon align, the Sun is on the other side of the Earth. In a total lunar eclipse the Moon traverses into the darkest part of the Earth's shadow, called the umbra. When the Moon is within the umbral shadow it will turn a reddish hue. Depending on the positioning of the Moon within the umbral shadow, lunar eclipses are sometimes called a "Blood Moon" because of this phenomenon. During a lunar eclipse, the Moon turns red because the small amount of sunlight that reaches the Moon is passing through Earth's atmosphere. The more dust or clouds present in Earth's atmosphere during the eclipse, the redder the Moon will appear. It's as if all the world's sunrises and sunsets are projected onto the Moon. Similarly, when we see the Sun setting after a fire that has filled our local atmosphere with ash and dust, as the Sun lowers on the horizon and is viewed through our atmosphere the Sun appears deep orange to red.

Technical Lunar Eclipse Information

Let's first look at the Total Lunar Eclipse of 1 BC, it was deep into umbral shadow of Earth, this would make it very dark, grayish color and dull at 99.99%-disc visibility.

The partial lunar eclipse of 4 BC was 99.98%-disc visibility is only one one-hundredth of a point difference. The only reason it is called a partial is because it is not 100%. The lunar eclipse of 4 BC would have been even more brilliant, being slightly closer to the edge of the umbral shadow not gray and dull but more orange. The argument that the lunar eclipse Josephus was was the 1 BC eclipse, merely because it was total at 99.99% vs partial at 99.98% when speaking of Lunar eclipses isn't significant. The fact that the 4 BC eclipse was partial, at the edge of the umbral shadow and brilliant, make it the more likely candidate.

The other point people attempt to make, is that both of these lunar eclipses were at night and people were in their houses and would not have seen them. During a lunar eclipse, the Sun must be opposite the Moon with the Earth in between. Those on the dark, night time side of the Earth can view the Moon as it is eclipsed by the Earth's shadow. We must realize where and when the 4 BC lunar eclipse took place. It was recorded as having been viewed in Jerusalem in March, one month before Passover. People lived, worked and slept both inside and on the rooftops of their homes. Many could have already been traveling with flocks on the road to Jerusalem, setting up pens to sell their lambs, or those wanting to set up booths to sell their wares, camping along the route, all in preparation for Passover. We cannot relate the event as though it occurred in the West; it was not New York, North Dakota, or Minnesota. In Jerusalem at that time, there was no light pollution, skyscrapers or Ponderosa Pines blocking their view, it was the dark skies of the Middle Eastern Desert.

Evidently the lunar eclipse of 4 BC was seen—Josephus recorded it! As he also explains the events that were taking place that night, when Herod had Matthias and his companions burned as punishment for the Golden Eagle event. This would have been a very long night, with many witnesses and protests.

Josephus writes *"Herod now fell into a distemper, and as he despaired of recovering, for **he was about the seventieth year of his age,** and with such discourses as this did these men excite the young men to this action; and a report being come to them that the king was dead, this was an addition to the wise men's persuasions; so, in the very middle of the day, they got upon the place, they pulled down the eagle,"*

Josephus' account of the Golden Eagle Event

"*And with such discourses as this did these men excite the young men to this action; and a report being come to them that the king was dead, this was an addition to the wise men's persuasions; so, in the very middle of the day, they got upon the place, they pulled down the eagle, and cut it into pieces with axes, while a great number of the people were in the temple. And now the king's captain, upon hearing what the undertaking was, and supposing it was a thing of a higher nature than it proved to be, came up thither, having a great band of soldiers with him, such as was sufficient to put a stop to the multitude of those who pulled down what was dedicated to God; so he fell upon them unexpectedly, and as they were upon this bold attempt, in a foolish presumption rather than a cautious circumspection, as is usual with the multitude, and while they were in disorder, and incautious of what was for their advantage; so he caught no fewer than forty of the young men, who had the courage to stay behind when the rest ran away, together with the authors of this bold attempt, Judas and Matthias, who thought it an ignominious thing to retire upon his approach, and led them to the king And when the king had ordered them to be bound, he sent them to Jericho, and called together the principal men among the Jews; and when they were come, he made them assemble in the theater, and because he could not himself stand, he lay upon a couch, and enumerated the many labors that he had long endured on their account, and his building of the temple, and what a vast charge that was to him;*"

"*But Herod deprived this Matthias of the high priesthood, and burnt the other Matthias, who had raised the sedition, with his companions, alive.* **And that very night there was an eclipse of the moon.**"

This eclipse of the Moon (which is the only eclipse of either of the luminaries mentioned in Josephus' entire work) is of the greatest consequence for the determination of the time for the death of Herod and Antipater, and for the birth and entire chronology of Jesus Christ. **It happened March 13th, in the** year of the Julian period 4710, and the 4th year before the Christian era. (4 BC)

Details in the account of Herod's illness, the Golden Eagle event and Herod's death

-Herod now fell into a distemper,
-and as he despaired of recovering, for he was about the **seventieth year of his age,**
-And with such discourses as this did these men excite the young men to this action; and a report being come to them that the king was dead, this was an addition to the wise men's persuasions; so, in the very middle of the day, they got upon the place, they pulled down the eagle,
When Herod called the principle Jewish men to address the punishment for the Golden Eagle event, he was very ill *"because he could not himself stand, he lay upon a couch,"*
"Herod's distemper greatly increased upon him after a severe manner,
"he had a difficulty of breathing, which was very loathsome, on account of the stench of his breath, and the quickness of its returns; he had also convulsions in all parts of his body, and having no longer the least hopes of recovering,"
"As he was giving these commands to his relations," at this very same time he was given permission by Augustus to have his son Antipater slain, Herod died 5 days later. *(Josephus Antiquities Book 17, 6:6-7:1)*

These brief excerpts from the accounts of his illness, Golden Eagle event, and his death provide a concise timeline and confirm Herod was **70 years old during the Lunar Eclipse** that occurred the night of the burning of the seditionists. **Herod was 70 years old during the lunar eclipse,** and he died shortly after. He was born in 74 BC; therefore, this confirms that **Herod died in 4 BC.**

Herod's will.

Herod's son Antipater, had been in line to be king after the death of his father, Herod. Due to Antipater's hatred and proven attempts to murder his father, along with many other treacherous deeds, Herod changed his will in favor of another son Archelaus.

"Hereupon Herod, who had formerly no affection nor good-will towards his son to restrain him, when he heard what the jailer said, he cried out, and beat his head, although he was at death's door, and raised himself upon his elbow, and sent for some of his guards, and commanded them to kill Antipater without tiny further delay, and to do it presently, and to bury him in an ignoble manner at Hyrcania." Josephus *Antiquities of the Jews XVII (17 7:1)*

"AND now Herod altered his testament upon the alteration of his mind; for he appointed Antipas, to whom he had before left the kingdom, to be tetrarch of Galilee and Perea, and granted the kingdom to Archelaus. He also gave Gaulonitis, and Trachonitis, and Paneas to Philip, who was his son, but own brother to Archelaus by the name of a tetrarchy; and bequeathed Jarnnia, and Ashdod, and Phasaelis to Salome his sister, with five hundred thousand [drachmae] of silver that was coined. He also made provision for all the rest of his kindred, by giving them sums of money and annual revenues, and so left them all in a wealthy condition. He bequeathed also to Caesar ten millions [of drachmae] of coined money, besides both vessels of gold and silver, and garments exceeding costly, to Julia, Caesar's wife; and to certain others, five millions. **When he had done these things, he died, the fifth day after he had caused Antipater to be slain; having reigned, since he had procured Antigonus to be slain, thirty-four years; but since he had been declared king by the Romans, thirty-seven."** Flavius Josephus, *Antiquities of the Jews 17 8:1*

Josephus wrote, Herod "died the fifth day after he had caused Antipater to be slain" History shows Herod's son Antipater was slain and died in 4 BC.

Herod was Dead
*"Now **Archelaus paid him so much respect, as to continue his mourning till the seventh day; for so many days are appointed for it by the law of our fathers.** And when he had given a treat to the multitude, and left off his motoring, he went up into the temple; he had also acclamations and praises given him, which way soever he went, every one striving with the rest who should appear to use the loudest acclamations. **So he ascended a high elevation made for him, and took his seat, in a throne made of gold,** and spake kindly to the multitude, and declared with what joy he received their acclamations, and the marks of the good-will they showed to him;"* Josephus Antiquities

As for the claim that there was not enough time between the death of Herod and Passover because his son Archelaus would have mourned 30 days, we see, as is recorded by Josephus, that Archelaus did not take 30 days to mourn his father Herod (and was criticized for not doing so), but observed mourning for only seven days. (*Shiva— Jewish mourning tradition: It refers to the period of mourning lasting seven days following the burial of a first-degree relative.*) Archelaus mourned only 7 days and took the throne soon after Passover, April 11, 4 BC.

"The attempts by a few historians to prove that Herod the Great died in some other year have not met with general acceptance. For example, W. E. Filmer ("The Chronology of the Reign of Herod the Great," JTS 17 [1966] 283-98) uses contorted arguments in an attempt to establish that Herod died instead in 1 BC. As Timothy D. Barnes points out very well ("The Date of Herod's Death", JTS 19 [1968] 204-9), Filmer's thesis collides with two major pieces of evidence: (1) Herod's successors all reckoned their reigns as beginning in 5-4 BC (2) The synchronisms with events datable in

the wider context of the history of the Roman Empire - synchronisms made possible by Josephus' narrative of the circumstances attending Herod's death - make 1 BC almost impossible to sustain."
John P Meier on the Year of Herod's Death 4 BC
NT History

History provides solid support for Herod's death in **4 BC**, not 1 BC or any other year. Herod's sons began their reign in 4 BC. There is no evidence of a Herodian coregency with any of Herod's sons. Archelaus and Antipas did not take their positions to rule until after the death of Antipater and **after** the death of Herod their father.

Herod's successors:
Herod Archelaus **4 BC** to AD 6 Ethnarch
Herod Antipas his reign in **4 BC** to AD 39, Tetrarch
Philip the Tetrarch began his reign in **4 BC**
Salome 1, Herod's sister. Upon the death of Herod the Great in **4 BC**, she was given a toparchy including the cities of Iamnia, Azotus, Phasaelis, and 5000 drachmae.

The Herodian family is important in New Testament history. Herod Antipas is mentioned in Mark 6:14, Matthew 14:1, Luke 3:1. Antipas was Tetratch of Galilee and Perea from 4 BC to AD 39. His son Herod Philip mentioned as Philip in Mark 6:17, Matthew 14:3, Luke 3:19.

Records show King Herod's reign ended in 4 BC, and it is also when his son Herod Antipas became Tetrarch over Galilee, and his son Philip became Tetrarch of Iturea, Trachonitis. Herod Archelaus (Matt 2:22), son of Herod the Great and Malthace of Samaria, brother of Herod Antipas, became Ethnarch of Samaria. He came to power after the death of his father, Herod the Great, in 4 BC and ruled over one-half of the dominion of his father. The date of death, **4 BC,** is confirmed by the dates of the beginning of the reign of Herod's successors.

Magi were warned in a dream about Herod. Scriptures and historical writings prove Herod's jealousy. He was notoriously paranoid and executed several of his own sons—**including** Alexander, Aristobulus, and Antipater—because he suspected them of plotting to overthrow him. He would kill anyone he **perceived** to be a threat to his kingship, including his **sons** and unrelated Jewish toddlers. They say Caesar Augustus remarked, "It is better to be one of Herod's pigs than one of his sons."

Matthew 2:12- *And having been warned in a dream not to go back to Herod, they returned to their own country by another way.*

Angel Appears to warn Joseph
Matthew 2:13-15 *Now when they had gone, behold, an angel of Adonai appeared to Joseph in a dream, saying, "Get up! Take the Child and His mother and flee to Egypt. Stay there until I tell you, for* **Herod is about to to search for the Child to kill Him."** *So he got up, took the Child and His mother during the night, and went to Egypt.* **He stayed there until Herod's death.**

Matthew 2:16 *Then when* **Herod** *saw that he had been tricked by the magi, he became furious. And he sent and killed all boys in Bethlehem and in all its surrounding area, from two years old and under, according the time he had determined from the magi.* (TLV)

Matthew 2:19 *But* **when Herod was dead,** *behold, an angel of the Lord appeareth in a dream to Joseph in Egypt, saying, "Arise, take the young Child and His mother, and go to the land of Israel, for those who sought the young Child's life are dead."* (NKJV)

Matthew 2:21-23 *And he arose and took the young child and his mother, and came into the land of Israel. But when he heard that* **Archelaus did reign in Judea in the place of his father Herod,** *he was afraid to go there; Being warned of God in a dream, he turned aside into the parts of Galilee. And he came and dwelt in a city called Nazareth, that it may be fulfilled which was spoken by the prophets, He shall be called a Nazarene.* (KJV)

JUDAEA, Herodian Kings
Herod Archelaus (4BC - 6AD)
Æ Prutah (2.72 grams) Jerusalem mint
grapes on vine/Military helmet

Archelaus 4 BC – AD 6

Roman procurator; treasurer of Augustus. "After Varus had returned to Antioch, between Passover and Pentecost of the year **4 BC** Sabinus arrived at Cæsarea, having been sent by Augustus to make an inventory of the estate left by **Herod on his death.**"

The consensus view of Herodian chronology and Herod the Great's Roman appointment as King of Judea in 40 BC, his taking of Jerusalem in the summer of 37 BC, and his death to 4 BC can be found in the article below.
Associates for Biblical Research
The Parthinian War Paradigm and The Reign of Herod the Great
Rick Lanser MDiv

Based on historical writings we know the reign of Archelaus began in 4 BC, Herod was dead at that time. Jesus was born before the death of Herod; we know the Magi visited Herod by following "His star" therefore Jesus was born before 4 BC. The date of Herod's death can only be March **4 BC** based on evidence.

Scriptural	check
Historical	check
Archaeological	check
Astronomical	check

Additional Evidence Supporting Herod's Death Date (Chapter 25)

Chapter 9

Caesar Augustus
and the Census

***Luke 2:1-4** Now in those days a decree went out from **Caesar Augustus,** that a census be taken of all the inhabited earth. This was the first census taken while Quirinius was governor of Syria. And all the people were on their way to register for the census, each to his own city. Now Joseph also went up from Galilee, from the city of Nazareth, to Judea, to the city of David which is called Bethlehem, because he was of the house and family of David,* (NASB)

Who was Caesar Augustus?
Historic Information Archaeological Evidence of His Acts.

Caesar Augustus was born Gaius Octavius September 23, 63 BC, adopted by his great-uncle Julius Caesar, he was named as the heir. He changed his name to Gaius Julius Caesar Octavius aka Octavian. Julius Caesar died in 44 BC. The senate awarded Octavian the name Augustus, (meaning "revered one"). From 27 BC – AD 14, Augustus was the first Roman Emperor. His stepson Tiberius was the second Roman Emperor; his co-reign began in AD 12, and his formal reign from AD 14 - AD 37, during the ministry of Jesus. Augustus collected numerous titles such as Pontifex Maximus (chief priest), Princeps (first citizen), Imperator (commander in chief) and divi filius (son of a god), which he took after the senate's deification of Julius Caesar.

When was the census of Caesar Augustus?
Do we have tangible archaeological evidence of a census of the Roman world by Caesar Augustus?
We have evidence from Caesar Augustus himself!

One of the most important inscriptions from the Roman era is the *Res Gestae (Acts or Achievements of Augustus.)* It is an autobiography in which Augustus describes his achievements. In his will, Augustus left instructions that upon his death these inscriptions were to be displayed in front of his mausoleum in Rome on two bronze pillars. Many copies of the Res Gestae text were carved in stone on monuments and temples throughout the Roman Empire. Of particular importance is the almost full copy which was written in the original Latin and a Greek translation was preserved on a temple to Augustus in Ancyra, there were other copies found at Apollonia and Antioch. The inscription in stone is on display in Yalvac Museum near Pisidian Antioch in Turkey. In his *Acts* Caesar Augustus describes taking a census of the Roman Empire on three different occasions.

Acts of Augustus

In my fifth consulship [29 BC] I increased the number of patricians on the instructions of the people and the senate. I revised the roll of the senate three times. In my sixth consulship with Marcus Agrippa as colleague [28 BC], I carried out a census of the people, and I performed a lustrum after a lapse of forty-two years; at that lustrum 4,063,000 Roman citizens were registered in the census roll. A second time in the consulship of Gaius Censorinus and Gaius Asinius (8 BC) I again performed the Lustrum alone with the Consular Imperium. In this Lustrum 4,233,000 Roman citizens were entered in the census roll. (registered). Thirdly I performed a lustrum with consular imperium, with Tiberius Caesar, my son, as colleague, in the consulship of Sextus Pompeius and Sextus Appuleius [AD 14]; at that lustrum 4,957,000 citizens were registered. By new laws passed on my proposal I brought back into use many exemplary practices of our ancestors which were disappearing in our time, and in many ways I myself transmitted exemplary practices to posterity for their imitation.

Credit: Todd Bolen | Bibleplaces.com
This *Res Gestae* (Acts of Augustus) inscription is on display in the Yalvac Museum near Pisidian Antioch in Turkey.

Archaeological evidence, a tangible historical record in the words of Caesar Augustus in his autobiography in stone. He said in his own words, that he conducted three censuses during his reign, the first was in 28 BC, the second census in 8 BC, and a third census in AD 14.

The first census of 28 BC being far too early and the date of the third census in AD 14 being far too late. Based on Scripture and archaeological evidence, the census date that is closest to the final days of the reign of Herod, is the second census conducted by Caesar Augustus in 8 BC. This 8 BC census is the only census that fits the timeline for the birth of Jesus— occuring during the reign of Herod and before the reign of his son Archelaus in 4 BC.

An objection raised has been that the Roman census was not to tax Jews only Roman citizens. However, Josephus records that the Jews were being taxed by the Romans with commands coming from Syria, as early as 44 BC. The task of raising the funds fell upon the Jewish rulers in power at the time. Josephus records: *"Cassius rode into Syria in order to take command of the army stationed there, and on the Jews, he placed a tax of 700 silver talents. Antipater gave the job of collecting this tax to his sons . . ."* Jewish Antiquities XIV 271

The Jews were required to pay taxes to the Romans, recorded in Roman history and also mentioned in Mark 12: 14-17 (NASB) *They came and said to Him, "Teacher, we know that You are truthful and do not care what anyone thinks; for You are not partial to anyone, but You teach the way of God in truth. Is it permissible to pay a poll-tax to Caesar, or not? Are we to pay, or not pay?" But He, knowing their hypocrisy, said to them, "Why are you testing Me? Bring Me a denarius to look at." And they brought one. And He said to them, "Whose image and inscription is this?" And they said to Him, "Caesar's." And Jesus said to them, "Pay to Caesar the things that are Caesar's, and to God the things that are God's." And they were utterly amazed at Him.*

"Another objection has been that Luke is in error. Josephus mentions a census which he associates with Quirinius' arrival in Syria to depose of Archelaus's money (ca. AD 6), which led to a tax revolt led by a man named Judas. Many critics believe that Luke has conflated these two events, mistakenly thinking that the census ordered by Caesar Augustus in AD 6 when he sent Quirinius into Syria, took place years earlier when Jesus was born. Two responses have been made that answer this objection. First, many scholars note that Luke was well- aware of the census of AD 6 and mentions it in Acts 5:37: "... *Judas the Galilean appeared in the days of the census and led a band of people in revolt. He too was killed, and all his followers were scattered."* This is why when Luke is writing about the census during the birth of Jesus, Luke explicitly states that "this was the *first* census that took place while Quirinius was governor of Syria" (Luke 2:2,)

"Another objection is Rome would not have required people to return to their ancestral homes, as described in the biblical account. However, surviving census information from Egypt, a Roman province at the time, indicate that this may have been the

common practice. For example, British Museum Papyrus 904 (ca. AD 104) contains a decree from the Prefect, Gaius Vibius Maximus, that, *"all those who for any cause whatsoever are residing outside of their provinces [are] to return to their own homes"* for a census."
Bible Archaeology Report
Bryan Windle
Caesar Augustus: An Archaeological Biography

    Josephus in his work *Antiquities of the Jews,* states, "When all the people of the Jews gave assurance of their good will to Caesar, and to the King's government; these very men [Pharisees] did not swear: being above six thousand." Some scholars have theorized that there was an empire-wide registration associated with this event. A census, or registration of some kind, would explain how Josephus knew there were 6000 pharisees who refused to take the oath to Caesar.
    A speculative idea has been that Caesar Augustus did not take any other census for himself or for other reasons, however as we can see below this was not the case. In the Res Gestae Acts of Caesar Augustus below it explains it was not always registration for taxes but for the "oaths" or "vows". Animal sacrifices were included, and Pharisees would likely not want to be involved in this type of oath taking and was a cause for revolt. The Jews often found cause to revolt.

*"The senate decreed that **vows** should be undertaken every fifth year by the consuls and priests for my health. In fulfilment of these vows games have frequently been celebrated in my lifetime, sometimes by the four most distinguished colleges of priests, sometimes by the consuls. Moreover, all the citizens, individually and on behalf of their towns, have unanimously and continuously offered prayers at all the pulvinaria for my health." Res Gestae of Caesar Augustus*

Pulvinaria are couches for the gods used in Roman religious rituals.

Both Scripture and archaeological evidence give us the date of 8 BC for the census in Luke where it is made very clear it was for **"all the inhabited world" to be registered,** (Luke 2:1). As we read in the Acts of Caesar Augustus, during the second census *"A second time in the consulship of Gaius Censorinus and Gaius Asinius (8 BC) I again performed the Lustrum alone with the Consular Imperium.* In this census it is recorded that 4,233 million were registered. This census date is not chosen based on what we would like to believe occurred during the time of the birth of Jesus. It is not based on our own traditions, stories, or latest ideas, but rather on the actual census written in stone in the **Acts of Caesar Augustus.** This extensive, far-reaching, world-wide census (not local) was taken preceding the death of Herod. The census of Caesar Augustus that fits Luke 2:1–3 was the census in **8 BC.**

The coin of Caesar Augustus with a "star" on it, is this the Bethlehem Star?
Julius Caesar and his adopted son Caesar Augustus were deified by the Romans.

We are fortunate the Romans were obsessed with taking notes and writing down information about everyone. Ancient writers tell us that there was a comet that appeared following the death of Augustus' father, Julius Caesar, in 44 BC. In July of that year, Caesar Augustus held a series of public games to honor the reign of his adopted father, Julius Caesar. The games established in Rome may have been similar to the Olympics. According to Roman historians, they noticed a comet hanging over the city of Rome during the games. The Roman historian Suetonius reports that during the course of the games, "a comet (stella crinita) shone throughout seven days in a row, rising at about the 11th hour, and it was believed that it was the soul of Caesar who had been taken up into heaven."

Credit: Classical Numismatic Group, Inc. | http://www.cngcoins.com | Wikimedia | CC BY-SA 3.0
A denarius issued by Caesar Augustus.

The coin reads *Divus Iulius* = Divine Julius

It was 44 BC when Julius Caesar was killed. Augustus commemorated his father by having the coin struck in his fathers honor. The comet on this coin was not the star of Bethlehem, nor did it appear near the time of the birth of Christ.

Many believed that the comet was the soul of Julius Caesar being divinized, the technical term that scholars use is "apotheosis", meaning to become a god. They believed it was the soul of Julius Caesar in the sky, that he had become a god and had been divinized. Hence, the commemorative coin was struck with the term "Divine Julius." It was a symbol of the *apotheosis* of Julius Caesar, Caesar becoming divine. The coin was a commemorative coin, similar to modern coins with an image of George Washington. Note on the coin: the words "Caesar Augustus" are on the side of the image, and "DIVVS IVLIV" (Divine Julius) on the side of the comet. Latin had no "J"; The letter "*I*" served as both vowel and consonant (pronounced "Yoo-lee-us")

With Julius Caesar being divine, Caesar Augustus was his son; **the Romans called Augustus the son of a god.** The comet on this coin was not the Star of Bethlehem, as some people are being misled to believe.

The Coin with Star and a Ram.
This coin, featuring a star and a ram, it is NOT the Star of Bethlehem. The coin below was minted many years after the birth of Jesus. The Romans were not interested in minting a coin that represented the deification of anyone other than one of their own kings or gods. Roman coins acted as a means of exchange and a tool for propaganda and symbols of their power.

Issuer: Antioch of Orontes
The [Antioch](#) mint at one time operated the largest number of *officinae* any Roman mint ever had operational. Antioch of Syria is located along the Orontes River in modern-day Turkey. At one time, this Hellenistic city was one of the largest in the Roman world. During the time of the New Testament, Antioch was a center of commerce and an important political power in the Roman Empire.

Antioch, Syria, AE Legate issue under Augustus. It is from the actian year 44, meaning it was struck between AD 12 to13.

ΑΤ and ΒΜ - ΜΗΤΡΟΠΟΛΕΩΝ ΑΝΤΙΟΧΕΩΝ, "people of the metropolis Antioch."
ΓΜ and ΔΜ - ΕΠΙ SILANOU ΑΝΤΙΟΧΕΩΝ, "reign of Silanus Antioch"
Antioch, under the Romans, AE20mm. Struck AD 13-14, reign of Augustus, Legate Silanus. Laureate head of Zeus right.
Minted in 13 AD.
Caesar Augustus was still alive when this coin was struck, the legate/governor of Syria at that time was Silanus.

Augustus used coins as perfect propaganda tools to enforce the new imperial cult. The coin was minted during the life of Caesar Augustus representing the divinity of his father Julius Caesar. Julius Caesar died in March 44 BC. The running ram looking back represents **Aries** also the month of March, the month in which Julius Caesar died, and the star which is shown is an eight-pronged comet, representing the comet that appeared during the deification of Julius Caesar. The head on the coin is the laureate head of **Zeus,** not only a god of Rome, but to the Romans, Zeus was the supreme god. As a divine king, his power was similar to that of the emperor's also divine. Under the Empire the leaders promoted Jupiter as a parallel to their own right to unchallenged and complete rule. The syncretism between Roman and Greek gods primarily took place during the Hellenistic period, which spans from around 323 BC to 31 BC. During this time, the Romans encountered Greek religion and began merging their gods with those of the Greeks. This process often comprised of equating Roman deities with their Greek counterparts, such as Jupiter with Zeus. This harmonizing of the gods helped promote cultural integration and governance over newly conquered territories.

It is most important to understand Roman mythology and the meaning of the coin, considering it was the Romans who minted the coin. The central myth of Aries is the Crysomallus, a fantastic flying golden ram created by Hermes (Mercury) to rescue the children of Phrixus and Helle from their wicked stepmother. After the task was complete, the ram was sacrificed to Zeus, and its golden fleece was hung in an oak in a grove sacred to Ares (Mars). March is named after the god of Mars; March is the month in which their divine Julius Caesar was killed. To the Romans, the Ram Aries represented determination and power. Augustus Caesar was determined, from the start, to avenge the death of his father, Julius Caesar, and to establish himself as the first Roman Emperor and "son of God."

Michael Molnar an astronomer and coin collector, has created an improbable story that is circulating and increasing the sale of this particular coin. Molnar's story began with the coin, "I did not set out to find an explanation of the Star of Bethlehem," he says. Looking at the coin, he decided to check the dates when Jupiter was in Aries, and found the dates of March 20 and April 17, 6 BC. He then referenced it against the years when he believes Jesus was born. Molnar is basing his conjecture on his own predetermined date of Jesus' birth; he starts with a coin and finds information that he forces to fit his theory. He skipped over Scripture, also historical, archaeological and astronomical pieces of data and evidence to support his theory are missing.

The bust on the coin is Zeus. It depicts a *comet* on the coin, not the planet Jupiter. It was minted from 16 to 20 years after the birth of Jesus. Some people attempt to suggest it was minted so people would look back on the birth of Jesus. At the time it was minted, no one but Mary and Joseph believed Jesus was anyone other than a poor Jewish teenager. The Romans minted coins for their own political purposes, deification, propaganda, and power. This particular coin is said to have been minted in Antioch Syria to celebrate their power and annexation of Judea and Samaria by the Romans, when Archelaus (son of Herod) was removed from his post by Caesar Augustus in AD 6.

There is a clear problem with Molnar's proposed date of March 20 (announcement) and April 17, 6 BC as the Star of Bethlehem. We must not ignore the most obvious and serious concern about these events, Jupiter **could not be seen!** Jupiter sets every night in March at sunset, this is not unique. On April 17th, Jupiter is barely visible for minutes, it was too close to the Sun. It could not be seen, it could not be seen over Bethlehem, there was nothing to see (as verified in Starry Night software).

This was a Roman coin, it is important to understand what Aries represented to the **Romans** at that time, Aries represents the fierce loyalty of the ram battling to protect those in need, the perfect propaganda piece for Augustus. Of more importance, is to understand what Aries meant to the Magi. Jupiter in Aries on

April 17, 6 BC, did not occur in the constellation Pisces, which represented Israel to the Magi (Babylonian Talmud). Jupiter was in Aries and Saturn was in a separate constellation Pisces, it also could not be seen. Jupiter and Saturn were <u>separating</u> quickly at a rate of nearly 3.5 degrees per month (after a previous conjunction having occurred the year before, in **7 BC**, as discussed in later chapters). However, Molnar went to great lengths to show that the Magi tracked the planets mathematically and were not dependent on visual verification, saying this was the reason that no one in Judea was aware that the star had occurred. Molnar's remedy for the fact that it was washed out by the Sun, and it couldn't be viewed was, that the Magi "were not dependent on visual verification." This is pure speculation and contrary to Scripture.

We must stick with Scriptures that pertain to the Magi and star event. Unfortunately what Molnar presents is not scriptural, based on **the words of the Magi** in Matthew 2:1-2 "For **we saw his star** in the **east**" later after leaving Herod they went their way and again viewed the star Matthew 2:9 "And behold **the star they had seen** in the east went on before them, until it came to rest over the place where the child was."

The above coin was not minted in remembrance of the Bethlehem Star, but a commemorative Roman coin struck by Caesar Augustus in honor of his father Julius Caesar. Nor was Jupiter in Aries (a common event and one the Romas couldn't see) on April 17^{th,} 6 BC the Star of Bethlehem. Jupiter setting for approximately 15 minutes, would not last long enough to be the Star of Bethlehem. It did not lead the Magi from the East to Jerusalem, nor rise and settle over Bethlehem.

Conclusion: Based on the evidence above, the census of Caesar Augustus in 8 BC provides the most conclusive timeline and motivation for Joseph and Mary's journey to Bethlehem.

Scripture	check
Historical writings	check
Archaeological evidence	check
Timeline confirmed	check

Chapter 10

Quirinius

Luke 2:1-2 *And it came to pass in those days that a decree went out from Caesar Augustus that all the world should be registered. This census first took place while Quirinius was **governing** Syria. So all went to be registered, everyone to his own city.* (NKJV)

Publius Sulpicius Quirinius (51 BC – AD 21) was a Roman general and governor who subdued at least two defiant and turbulent regions of the Roman Empire for Caesar Augustus. Born near Rome, Quirinius rose through military and political ranks to become a key figure in imperial expansion. At age 37 after Quirinius defeated a North African tribe the Marmaridae, that was plundering in Cyrenaica (Libya) in 14 BC, Caesar Augustus made him Consul in 12 BC and dispatched him to wage war against the Homonadenses, a tribe in the mountains of Cilicia (Turkey) which is adjacent to Syria. Quirinius was successful in leading his military campaign and defeated the Homonadenses and was governing Syria during the time of Luke 2:2's census (Tacitcus, *Annales* 3.48).

Consul: In the Roman Republic, either of the two who held highest of the ordinary magistracies in the ancient Roman Republic. After the fall of the kings, the consulship preserved regal power in a qualified form. Absolute authority was expressed in the Consul's imperium. The position of Consul was often the high point of a Roman politician's career. After he left office, he remained a member of the Senate and would most often be rewarded for his service and named **governor** of one of the Roman provinces, a pro-consul.

Who was Quirinius and what were his responsibilities?
Publius Sulpicius Quirinius 51 BC - 21 AD
Senator, Consul, Legate, Duumvir he held many roles.
Named by Josephus, Tacitus, Pliny the younger, Suetonius, Cassius Deo, Stravo and Caesar Augustus himself.

Luke 2:1"A decree went out from Caesar Augustus **"to register"** (Gk a*pographesthai*) the "world" (Gk *oikoumene)"* signifying all the inhabited earth during New Testament time referenced to the Roman Empire.
Luke 2:2"This "first" (Gk *prote*) registration took place when Quirinius **"was governing"** (Gk *Hegemoneuontos)* Syria.
Luke 2:3 "So all went "to be registered" (Gk *apographesthai)* each to his own city."

When was Quirinius "governing"?
    While some translate prote to mean "before" Quirinius was governor. Luke makes a very clear point by mentioning "this was the **first** census". Either way you look at it, Luke was fully informed of the first census, and the registration and revolt mentioned in Acts 5:37

The census in the Gospel of Luke vs the census in Acts 5:37

    Regardless of exegetical, historical and archaeological indications to the contrary, some scholars have determined Luke's writing of the later census in Acts 5:37 to be the one that took place when Jesus was born. In researching their claims, I have gone down their rabbit holes and found that in their approach they had already drawn the red line and determined in their own mind's certain dates for the birth and death of Jesus. Starting from the wrong point, that point being their erroneous presumption for the birth date of Jesus. Then they sought out bits of information in an attempt to support their theory that "Luke got it wrong" regarding the census in Luke 2:1-3, and that this census was the census Luke mentioned in Acts 5:37 which took place at a much later

date. Therefore, their conclusion that Jesus was born later, relies on artificially moving Herod's death to 1 BC. forcing the information to fit their own predetermined ideas.
(Objections explained in detail in *The Bethlehem Star* by Dr David Hughes, Astronomy Professor Department of Physics and Astronomy University Sheffield.)

The confusion arises when Luke's reference to a later census in Acts 5:37 is conflated with the Nativity account. Emil Schurer, in *A History of the Jewish People at the time of Jesus Christ*, concluded, "there is no alternative but to recognize that the evangelist based his statement on uncertain historical information". Schurer's skepticism, however, reflects his own lack of faith in Scripture's reliability rather than any inaccuracy on Luke's part. As a meticulous historian, (Luke 1:1-4), Luke relied on eyewitnesses who had lived during the events he recorded. Their testimony outweighs modern doubts rooted in incomplete archaeological or textual evidence. Luke wrote an *"orderly account"* with *"perfect understanding"* he mentions two censuses, and it was during the first census that Jesus was born. It would be very unlikely for such a conscientious historian to make a mistake in his timeline of events.

Luke 1:1-3 *"Many have undertaken to draw up an account of the things that have been fulfilled among us, just as those who from the beginning were eyewitnesses and ministers of the word delivered them to us, it seemed good to me also,* **having had perfect understanding of all things from the very first**, *to write to you an orderly account, most excellent Theophilus, that you may know the certainty of those things in which you were instructed."*

In this book, I prioritize Scripture as the inerrant foundation of truth, the definitive witness to these events. By presenting corroborating historical, astronomical, and archaeological evidence, I allow the data to affirm and the evidence will speak for itself.

Historical Writings and Explanations Prove That Luke Was Entirely Accurate.

Quirinius was a special Lieutenant of Augustus, who conducted the war against the Homonadenses, while Varus administered the ordinary affairs of Syria. The duties of Quirinius might be described by the Latin term *dux*, meaning a military or administrative leader. In Greek, the equivalent term is *hegemon*, which Luke uses correctly to convey the same sense of authority and responsibility. Luke rightly describes the authority of Quirinius by the words "holding the Hegemonia of Syria". The exposition on *Bible Hub.com* regarding Quirinius the Governor of Syria, details his titles, administrative rolls and chronology, resolving any ambiguity about Luke 2:2's reference. Quirinius was exercising this delegated Hegemonia over the armies of the Province Syria, and it seems quite in keeping with Luke's explanation to say that he held the Hegemonia of Syria. This position Quirinius held in 7-6 BC. This interim military governorship (distinct from his later senatorial governorship in AD 6) explains how he could oversee Judea's census *before* Herod's death in 4 BC, harmonizing Scripture with extrabiblical evidence."

Archaeological Evidence for the census under Quirinius on the tombstone writings from his servant.

The discovery of a tombstone in Beirut known as the "Aemelius Secundus Inscription" The tombstone of Q. Aemelius Secundus, who conducted a census for the Legate Quirinius in Apamea Syria.

The census that Quirinius conducted in Syria has been confirmed by an inscription on the stele purchased in Beirut in 1674 and brought to Venice commemorating a Roman officer who had served under Quirinius stating additional achievements "By order of the same Quirinius I took a census of the city of Apamea". The inscription clearly implies that the Homonadenses were conquered in Quirinius' **first** Syrian administration.

There are plenty of instances where people claim there is no evidence, then a discovery is made, proving the Bible was completely accurate and affirming its authority and its reliability.

Bible skeptics claimed that Pontius Pilate, the Roman governor who sentenced Jesus to be crucified was a fictional character because there was no secular confirmation. Then came the discovery of Pontius Pilate's name engraved on a limestone fragment (pictured previously), in the ruins of the Roman stadium in Caesarea in 1961. Although many believe we don't need secular sources to support the infallible Word of God, we continue to see the Bible be corroborated by archaeology and advances in technology and other sciences.

Timeline
Quirinius was in office from 12 BC to AD 6

The possible tombstone of Quirinius himself.
The inscription on a tombstone discovered near Trivoli Italy its owner was "Twice Legate" of Augustus in Syria. The name of the owner is not attached to the tombstone due to its fragmentation, but it proves that a person can be Legate/Governor twice, and this person was Governor of Syria under Caesar Augustus more than once.

This inscription is from a tombstone discovered near Trivoli, Italy. It's owner was once "Twice Legate" of Augustus in Syria. Photo credit: Todd Bolen, BiblePlaces.com NOTE: This Photo is part of an excellent resource called the Photo Companion to the Bible – https://www.bibleplaces.com/luke-photo-companion-to-the-bible/

This tombstone confirms the high probability Quirinius was Legate/Governor twice. Luke said when "He was governing Syria" that is when he was overseeing a census.

Another claim is that Luke's reference to the census represents an error and Joseph and Mary could have been counted in Nazareth, as there was no requirement to return to Bethlehem. That is contrary to Scripture as seen in Luke 2:3-4 and it is supported by secular sources (mentioned earlier) in the British Museum Papyrus 904 (ca. AD 104) which contains a decree from the Prefect, Gaius Vibius Maximus, that, *"all those who for any cause whatsoever are residing outside of their provinces [are] to return to their own homes"* for a census." Also, Josephus provides most of the information, in addition we know that Caesar Augustus' rule was marked by a significant increase in census activity.

Luke 2:3-4 *And all the people were on their way to register for the census, **each to his own city.** Now Joseph also went up from Galilee, from the city of Nazareth, to Judea, to the city of David which is called Bethlehem, because he was of the house and family of David.*

Justin Martyr wrote:
Hear also in what part of the earth he was to be born, as another prophet, Micah, foretold. He said, "And you Bethlehem, land of Judah, are by no means the least among the rulers of Judah; for out of you will come a Ruler who will shepherd my people." This is a village in the land of the Jews, thirty-five stadia from Jerusalem, in which Jesus Christ was born, as you can learn from the census which was taken under **Quirinius, who was your first procurator in Judea.** *737*
Procurator: title of the governors

Based on the evidence presented above, Quirinius was "governing" at the time of the 8 BC census.

Scripture   check
Historical writings   check
Archaeological evidence check

Disclaimer: The information presented in the following chapters, is a study drawn from historical, Biblical and Hebraic understanding of the "signs" which God placed in the heavens at creation (Genesis 1:14) God's people were instructed not to be "dismayed at the signs in the heavens" as the Gentiles were. Worship of the Sun, the Moon and stars were all strictly forbidden in the Bible.

Chapter 11

Wise Men Follow His Star

What does Scripture say?
Matthew 2: 1-11
Now after Jesus was born in Bethlehem of Judea, in the days of King Herod, the magi from the east (anatolon) came to Jerusalem, saying, "Where is the one who has been born King of the Jews? For we saw His star in its rising (anatole) and have come to worship Him."

Matthew 2:1 Greek *Magoi apo anatolon paregenonto eir Hierosolyma* – Magi **from the east** arrived in Jerusalem. Greek words *Magoi **apo anotolon*** = Magi **from the east**

Matthew 2:2 Greek *eidomen gar autou ton astera **en te anatole*** – *we saw his star* (in the east) *in it's rising and have come to worship him."* Greek words *ton astera **en te anatole*** = the star **in it's rising,**

Matthew 2:1-11 *"When King Herod heard, he was troubled, and all Jerusalem with him. And when he had called together all the ruling Kohanim/Priests and Torah scholars, he began to inquire of them where the Messiah was to be born. So they told him, "In Bethlehem of Judea, for so it has been written by the prophet: "And you, Bethlehem, land of Judah, are by no means least among the rulers of Judah; For out of you shall come a ruler who will shepherd My people Israel."*
"Then Herod secretly called the magi and determined from them the exact time the star had appeared. And he sent them to Bethlehem and said, "Go and search carefully for the Child. And when you have found Him, bring word back to me so that I may come and worship Him as well."

"After listening to the king, they went their way. And behold, the star they had seen in the east went on before them, until it came rest over the place where the child was. When they saw the star, they rejoiced exceedingly with great gladness. And when they came into the house, they saw the Child with His mother Miriam; and they fell down and worshiped Him. Then opening their treasures, they presented to Him gifts of gold, frankincense, and myrrh."

Was the Bethlehem Star a Predictable Event?
The Bethlehem Star was a sign in the heavens; it had an appointment previously scheduled on God's astronomical clock. It was an astronomical event, and only the astronomers of the time understood the meaning and trajectory of "His Star."

Galileo understood that the Sun was the center of the Solar system, not the planet Earth, and therefore he was placed under house arrest by the Church. Due to the teachings of Ptolemy and Aristotle, the Western world thought humans were located at the center of the universe. In 1542, twenty-two years before Galileo was born, Nicolaus Copernicus published a heliocentric system. Galileo's heliocentric view was contrary to Ptolemy and Aristotle but not contrary to the Bible. Galileo did not believe the Bible was explaining the physics of the heavens. He quoted a cardinal saying **the Bible teaches "how to go to heaven not how the heavens go."** The pope turned against him, believing he was being mocked, and Galileo was forced to spend the remainder of his life under house arrest. But as time went on the works of Kepler and Newton, along with the advancement of technology his claim was proven to be true.

For heliocentric objectors, I kindly suggest giving consideration to these five words: "The Flight of the Voyager". As close relative of a pilot, flight instructor, engineer who played an important role in the success of that mission, I offer a suggestion: study the flight of the Voyager.

Why didn't the Religious Leaders see the Star?
Why didn't the religious leaders in Jerusalem recognize the star if it was supernatural and magnificent and looked like the star on a Christmas card? They saw nothing out of the ordinary, the star itself was not unusual, not at the time the Magi arrived in Jerusalem. Herod called the religious leaders into his court to ask about the prophecy of the "King of the Jews" being born, also because the Magi were inquiring around Jerusalem, saying, "Where is the One who has been born King of the Jews? For we saw *His Star* in the East and have come to worship Him."

One of the Church Fathers, Origen, in his *Contra Celsum* (not Scripture), went so far as to claim that the Magi were communing with evil spirits. Due to this and similar erroneous yet influential comments, a fear of the heavens persists among many Christians today. Like other signs that the religious leaders ignored, signs for which Jesus admonished them, they refused to consider the astronomical evidence. Out of fear and stubborn adherence to their traditions, they were oblivious to God's great clock pointing to the appointed time of His promise, and they missed their own Messiah. The religious leaders focused on the Law; the Magi watched.

Chapter 12

The Magi

The word Matthew uses, *magi* (in English) comes from the Greek *magoi,* correctly translated as magi in Matt 2:1 is a plural proper noun referring to people from the Ancient Near East, ancient Media and Persian astronomers or scholars and translated in the authorized version of the Bible as **"wise men"**, the New English Bible as "astrologers". The translation "wise men" is a very broad term, there are many wise men, but it doesn't tell us who they were, their proper title *magoi* does, the Magi were a specific group. The name Magi became associated with the hereditary priesthood. Because of their combined knowledge of science, agriculture, mathematics and history, their influence continued to grow until they became the most prominent and powerful group of advisors in the Babylonian and Medo-Persian empire.

In Matthew's account of the Magi, there is no mention of sinful false practices, the Magi were spoken of in an admirable light. Characterizing the Magi as pagan evil magicians because they were astronomers/astrologers is equivalent to the Church calling Mary Magdalene a prostitute just because she was a woman. Without accurate information, an association is made without evidence, nowhere in the Scriptures is Mary Magdalene identified as a prostitute. As we know by studying Scripture it clearly states Mary Magdalene was delivered from seven demons and became a faithful disciple of Christ (Luke 8:2-3).

It is evident from Scripture that the Magi were intellectually alert and possessed religious perceptivity; they were correct in their observations, having seen the revelation of the King of the Jews in nature!

Some have argued, the reason for the Magi's interest was to satisfy mythological or political propaganda, since everyone desired global peace. Nevertheless, the Magi attained their goal, having found the **long-awaited Jewish Messiah** through astronomy and astrology and they worshiped Him.

The Magi, the wise men of Babylon.
Daniel 2:48-49 *Then the king promoted Daniel and lavished on him many marvelous gifts and made him ruler over the whole province of Babylon and chief over all the wise men of Babylon. Moreover, at Daniel's request the king appointed Shadrach, Meshach and Abed-nego over the administration of the province of Babylon, while Daniel remained in the royal court.*

Daniel, Shadrach, Meshach and Abed-nego, while living in captivity in Babylon, were not influenced by the religion of Babylon and made their stand against it, even under the threat of death, (Daniel 13:12-30) they still remained wise men in the courts of the Babylonian king. There were also many descendants of the Jews and others who did not succumb to the prominent religion of the land, and there isn't any evidence that the Magi were gentiles. In Matthew the Magi make it very clear they came to find "the King of the Jews and worship Him". Matthew 2:2, *"Saying, where is he that is born King of the Jews? for we have seen his star in the east, and **are come to worship him**."* 11-12 *"And when they came into the house they saw the Child with His mother Mary; and they **fell down and worshiped Him**. Then, opening their treasures, they presented to Him, gifts of gold, frankincense, and myrrh. And **having been warned in a dream** not to go back to Herod, they returned to their own country by another way."* The Lord warned them in a dream! We see a similar situation with Joseph, another admirable person and the earthly father of Jesus, who was told in a dream to bring the child Jesus back to Israel. Matthew 2:22 *"Then after being warned in a dream, he withdrew to the region of Galilee."* In Matthew the Magi were portrayed as laudable and exemplary models, there was not a derogatory word, no condemnation of false practice, but very wise men who understood Psalm 19:2 *Day after day they pour forth speech; night after night they reveal knowledge.*

We don't know exactly what all the Magi believed. Many assume that they worshiped the stars (Astro**latry**) because of the region the Magi came from, presumably Babylon, where the worship of false gods was prevalent. Nevertheless, according to the actions of the Magi in Matthew 2:2, 10-12, they did not. It is highly likely they were monotheistic and followed the teachings of Daniel, as we will see in the paragraphs below.

Babylonian Captivity

During the reign of King Nebuchadnezzar II (607 BC), King Jehoiakim was forced into submission, and the Jews were taken into captivity to Babylon (2 Kings 24:1). During this time, Nebuchadnezzar took many of the finest and brightest young men from each city in Judah captive, including Daniel, Hananiah (Shadrach), Mishael (Meshach) and Azariah (Abed-nego). Babylon fell to Cyrus the Persian (539 BC), which led to the return of the Jewish exiles to their homeland. Although many Jews returned to their homeland (Ezra 2:64), after being in captivity for 70 years, a large group of Jews remained in Babylon. What we read in the book of Daniel confirms that the Magi were highly sought-after important members and advisors of the royal court.

During the Babylonian and Medo-Persian empire Daniel had a distinguished and influential position among the Magi. It was likely that the Magi in Matthew 2:1 were inspired by the prophet Daniel, they were men of science seeking the Jewish Messianic King. Influenced by the teachings of their most revered teacher, Daniel, they would have been encouraged in their understanding of monotheism and emboldened with the expectation of a coming Jewish King. With anticipation, they would search for and recognize a specific planetary conjunction, understanding its association with the birth of the *King of the Jews.*

The location of one's origin does not always dictate one's beliefs. Abraham came from Ur, but he did not worship their false gods. Today, many people come from different countries where the majority of the population worships other gods. Yet, some individuals, regardless of cultural influence, recognize and believe in the God of the Bible. Are there no Christians in India or in China? Of course there are. We only assume that the Magi, because of their origin, were pagan sinners. But the overwhelming evidence shows, they were likely influenced by Daniel and were men of science, astronomers and astrologers in the true sense of the word ("ology" meaning "the study of"). The Babylonians were world leaders in the knowledge of astronomy and astrology.

On the other hand, the religious leaders in Jerusalem clung to their traditions and self-righteous interpretations, they were not looking, not watching. Ancient Rabbi's believed in astrology and references to its insights are found throughout the Babylonia Talmud, where the influence of the planets and constellations are described. The religious leaders in Jerusalem avoided understanding astronomical events for fear that they would be associated with Gentile sinners and pagans. They were also well aware of the Scripture Jeremiah 10:2 *"Do not learn the ways of the nations or be terrified by signs in the heavens, though the nations are terrified by them.* Although they believed astrology affected their lives, they also believed that if a Jew was observant of Torah, it would override any influence of astrology. *(Babylonian Talmud)*

The Babylonian Captivity and Its Consequences

"In 587 BC Nebuchadnezzar besieged and took Jerusalem, allowed his troops to plunder parts of it, razed to the ground the temple that had stood for almost four hundred years, abolished the kingdom of Judah and carried several thousand inhabitants of Jerusalem off to destinations east of the Euphrates River."

The Babylonian Captivity can "be termed the beginning of Judaism in Mesopotamia." "From the first century until the end of antiquity many more Judeans lived in Mesopotamia than lived in Judea. And it was in Mesopotamia that rabbinic scholars produced the "Babylonian" Talmud, which has been normative for rabbinic Judaism ever since."

"Their worship of Adonai remained fervent, but they could no longer worship him in the traditional way. They had recently been instructed by the Book of Deuteronomy that only at the Jerusalem temple were sacrifices pleasing to the Lord, and so their worship in Babylonia could include no sacrifice."

"The importance of obedience to the Torah among Judahites of Mesopotamia is shown clearly in the books of Ezra and Nehemiah. Both of these men were born in Mesopotamia and spent most of their lives there, but in the prime of life were appointed by the Persian king to undertake a mission from Mesopotamia to Judea. Ezra was a *sofer,* a scribe learned in the Torah of the Lord....

Because it was in Mesopotamia that the Torah first became central for Judahites, the Mesopotamian legacy to Pharisaic and Rabbinic Judaism was enormous, but the legacy did not end with Ezra and Nehemiah. Late in the first century BC, Hillel came from Mesopotamia to Jerusalem and established there the Beth Hillel ("House of Hillel") for instruction in the written and oral Torah. The first attempts to establish and canonize the Hebrew text of the Pentateuch were made in Mesopotamia at or shortly before Hillel's time, and meticulous study of both the Law and the Prophets continued there all through antiquity. The Babylonian Talmud, completed ca. 500 AD, was in fact the culmination of more than a thousand years of Judaism in this a strange land."

The Judahite community in Mesopotamia not only survived but grew remarkably.

"In 539 BC, the word of the Lord had been fulfilled: Cyrus had conquered Babylon and had put an end to the Chaldean empire. Cyrus dealt kindly with the Babylonians, who welcomed him into the city with palm branches. Neither little children nor their parents were slaughtered. Throughout the two hundred years of Persian domination Babylon continued to be one of the great cities of the ancient world and served as the winter capital of the Persian empire. Although the city of Babylon suffered little damage in 539 BC, the Babylonians did lose their empire to

Cyrus, and tribute from the Fertile Crescent **no longer** poured into the great temple of Marduk. Men and women alike would need to follow the dietary laws of the Torah when they joined a Judean congregation, they would no longer be permitted to work on the Sabbath, and of course they would henceforth not participate in the holidays for the image gods that the rest of the city celebrated. Despite these handicaps, the attractions of Judean monolatry and of the Judean community prevailed. From the sixth century BC to at least the second century AD, the expansion of Judaism in Mesopotamia was remarkable. From the second century onward Judaism had to compete with Christianity, and then with Mazdaism and Manichaeism, and its growth slowed. Nevertheless, Judaism remained vigorous in Mesopotamia. It was there that two of the most important rabbinic academies were located - in the cities of Sura and Pumbeditha - and that the "Babylonian" Talmud was compiled."
Source and quotes from
The Babylonian Captivity and Its Consequences
By Robert Drews
Professor of Classical Studies Emeritus at Vanderbilt University.

The Magi knew what the star was, knew what it meant, how did they know?

During the time of Matthew, "the East" was a very vague term. Origen thought that "the East" referred to all the countries mentioned by the historian Josephus by that name, an area stretching approximately from Aleppo in the northwest to the modern city of Mosul in Iraq on the Parthian border. "It seems that, east of Palestine, only the ancient countries of Media, Persia, Assyria, and Babylonia had a Magian priesthood at that time."

Between 220 BC and AD 75, Babylonian astronomy had advanced so far, that all significant phenomena involving the five visible planets and the Moon, could accurately be computed in advance by astronomers and astrologers. This is demonstrated in the many Babylonian astronomical almanacs that have survived from this period. The Babylonian texts were prepared a year in advance and provide a month-by-month account of what would be seen in the night sky. The data include lunar and solar eclipses, solstices and equinoxes, the first and last dates when stars would be visible in the night sky.

How the Magi Viewed the Planets and Constellations.

The first mentions of the constellations moving along the path of the Sun, the Moon and planets belongs to the Babylonians having observed them in 1000 BC. Although at that time they nearly corresponded with the twelve zodiac positions in the sky, the names that were given to the constellations are the same as the zodiac signs, translated through the works of Greek astronomers and astrologers carry over to western or modern astronomy and astrology today.

The solar system within the constellations *Mazzaroth* is God's analog clock in the sky.

Book of Job 38:31-32 (God speaking to Job):
"Can you bind the chains of the Mazzaroth, or loose the cords of Orion? Can you lead forth the Mazzaroth in their season, or can you guide the Bear with its children?"

Meaning and Interpretation: The context is key. God overwhelms Job by listing the wonders of creation that are far beyond human control. The "chains" or "sweet influences" (in some translations) of the Mazzaroth refer to God's sovereign power to bring forth the constellations in their appointed seasons and maintain their order throughout the year.

Connection to the Zodiac: While the word "Zodiac" comes from the Greek *zōdiakos kyklos* ("circle of animals"), the concept of a path in the sky along which the sun, moon, and planets move, divided into constellations, was also known in the ancient Near East. The Hebrew *Mazzaroth* refers to this same concept: the band of constellations along the ecliptic path.

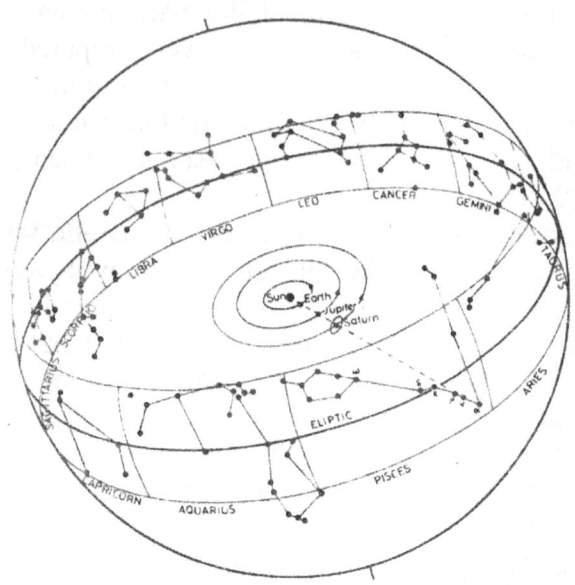

God's Analog Clock in the Sky

The Babylonian Astronomical Diaries, housed in the British Museum, contain records of astronomical observations in Babylonian cuneiform. The many fascinating artifacts tell a story of a civilization who believed their lives were interwoven with the stars and planets and constellations. They considered Jupiter the superior planet (wandering star). The Babylonian astrologers did not have the aid of telescopes and would have seen Saturn as we see it today with the naked eye; but in a dark night sky of the desert with no light pollution, they understood the paths of the wandering stars traversing the night sky. The times in which a planet came close to its planetary neighbors in the same constellation was significant and was recorded in the Babylonian astrological tablets.

What is remarkably different about the Babylonian Magi is, they viewed the stars and planets as prominent informants of the meaning of events in their daily lives. As is evident in the Talmud, they indeed felt very connected to the universe in ways which only they could understand.

From the Jewish perspective, the Zodiac signs were positioned at the time of creation, and their influence is described in the Torah. The Talmudic Sages go into great detail to describe the progression of the creation of the zodiac signs and their corresponding influence on Mankind (Pesikta, ch. 4). The entire wisdom of the Torah, including the knowledge of the zodiac, was taught by God to Adam and his descendants. Although it was eventually perverted and incorporated by most of humanity into idol worship, a select few such as Abraham retained its true meaning.

Job 38:32 *Can you bring out Mazzaroth (constellations) in its season? Or can you guide the Great Bear with its cubs?*

The ancient rabbis lived among peoples who believed in astrology and references to its insights are found throughout the Babylonia Talmud. Among the most famous is this line from Genesis Rabbah 10, which was written during the Talmudic period: "Rabbi Shimon said: "There is not a single blade of grass that does not have a constellation in the firmament that strikes it and says to it: 'Grow.'" In Moed Katan 28a, Rava states that one's lifespan, children and sustenance depend not on merit, but on mazala, the Aramaic word for constellation.

Why was Pisces Significant to the Magi?

The miracle of Purim during the time of Esther, Mordechai, and the unity between the Jews in Shushan, in Persia occurred during Pisces. This and unified effort, caused the miracle that prevented the total extermination of the Israelites in Persia under Ahasuerus' rule.

One of the earliest, ancient Jewish texts called Sefer Yetzira correlates the zodiac sign Dagim/Pisces as the hidden miracles of God which bring **salvation** and **joy to the Jewish people**.

It is interesting the ancient words in the Sefer Yetzira are also the words spoken by the Angel who visited the shepherds.

Luke 2:10-11 (NIV) *But the angel said to them, "Do not be afraid. I bring you good news that will cause **great joy for all the people**. Today in the town of David a **Savior** has been born to you; he is the Messiah, the Lord.*

Pisces Meaning

The Hebrew word *Dagim* is represented by the 10th letter of the Hebrew alphabet Yod.

      Yod   Yod

Modern Hebrew   Paleo Hebrew

Yod is a representation of an outstretched hand implying receiving something or recovering something. To help, hand, arm of the Lord, work, heal, power, authority. His hand is the ultimate authority in the land. Number 10 relating to the recovery of the lost sheep of Israel. *But He answered them and said "I am not sent but to the lost sheep of the house of Israel."*— Matthew 15:24

Saturn and the Jews
November 10, 2017
by Shlomo Sela

"Underlying the well-known link between Saturday (Shabbat in Hebrew) and Saturn (Shabbetai in Hebrew) is the reference to Saturn as the planet in charge of the Jews. Behind the link between Saturn and Saturday, is the astrological theory that assigns the seven planets in succession, beginning with the Sun and following the order of their orbs, to the 24 hours of the day and to the seven days of the week. Prominent Roman historians such as Tacitus (56–120 CE) and Cassius Dio (ca. 155–after 229), as well as Church fathers like Augustine (354–430), acknowledged a special link between Saturn and Saturday, the holiest day of the week for the Jews. **That Jewish society of the Talmudic period recognized the same association** is shown by the fact that the Babylonian Talmud (Shabbat156a) refers to Saturn as Shabbetai, i.e., the star of Shabbat (Saturday)."

How the Jews Viewed Saturn (in the Talmud 156a)

*"One who was born on Shabbat will die on Shabbat, because they desecrated the great day of Shabbat on his behalf. Rava bar Rav Sheila said: And he will be called a person of great sanctity because he was born on the sacred day of Shabbat. Rabbi Ḥanina said to his students who heard all this: Go and tell the son of Leiva'i, Rabbi Yehoshua ben Levi: It is not the constellation of the day of the week that determines a person's nature; rather, it is the constellation of the hour that determines his nature. He who is born under the influence of Saturn will be a man whose plans will be frustrated. Others say: He who is born under the influence of Jupiter (Tzedek) will be a right-doing man [tzadkan] R. Nahman b. Isaac observed: Right-doing in good deeds."*Babylonian Talmud: Tractate Shabbath 156a

Tzedeq, Jupiter Meaning: "righteousness", as Jupiter is the embodiment of divine influx.

Shabtai, Saturn Meaning: "of Shabbat" The planet represents the Jews.

Astronomy and Judaism

Rabbi Shimon ben Pazi, reported that Rabbi Yehoshua Ben Levie said in the name of Bar Kappara: Anyone who knows how to calculate astronomical seasons and the movement of constellations and does not do so, the verse says about him: "They do not take notice of the work of God, and they do not see His handiwork" (Isaiah 5:12).

"Astral beliefs were prevalent among the nations that surrounded the ancient Israelites. The Mesopotamian religions were completely astral, where each deity had a specific astronomical manifestation, and the motion of heavenly bodies was considered a practical means of the gods to communicate with Man. The Jews oscillated among adopting characteristics of the surrounding beliefs of the Canaanites, the Phoenicians and their likes, the strong Assyrian influence from Mesopotamia, and their own unique monotheistic culture" (Ness 1990, McKay 1973). In this context, the main use of astronomy was for calendrical purposes, using methods developed by the Mesopotamians. However, astrology in its simplest form also infiltrated strongly to Jewish culture.

Jupiter: Jupiter's Hebrew name is (tzedek), whose usual meaning is 'justice'. The Arab name is (almushtari), which means 'the buyer' or 'the owner'. A possible hint to a connection between these two seemingly different etymologies is mentioned in one tradition, according to which Jupiter shined all night during Abraham's fight with Chardolaomer, king of Eiloam. In the biblical text it is not mentioned, but it is mentioned that 'Malkitzedek" king of Shalem (Jerusalem) blessed Abraham. It is said that Malkitzedek is a priest of the god 'El Elyon', who was described in an Ugaritic text as 'buyer (owner?) of the heaven and earth'. 4.10.

Saturn: The current Hebrew name of Saturn is (*shabtai*), and one can easily hear an allusion to– Shabbat, the holy day of rest, Saturday. This name appears first in the Talmud. It seems that at some point during the Talmud period the Jews adopted the planetary week, where Saturn rules Saturday, and thus the Hebrew name actually means 'Saturday's planet'. There is one appearance in the Bible of another name, with unclear vocalization, which may read 'kayun', or 'kiwan' (Schiaparelli 1905). The source is probably Akkadian, where *'kayamanu'* means 'the steady one', which is similar to Hebrew *qayam* ('exists'), or Arabic *kawn* ('existence'). A relic to this ancient word exists in modern Persian, where Saturn is named *'kayvon'*. It may well be that the name was chosen to refer to the fact that Saturn is the most 'stable' planet, i.e., it changes its position relatively slowly. As for the Arab name *Zukhal,* I suggest its meaning may be related to the same root in Hebrew, where it means 'to crawl', 'to advance very slowly."

Hebrew Names of the Planets: Shay Zucker Department of Geophysics and Planetary Sciences, Raymond and Beverly Sackler Faculty of Exact Sciences, Tel Aviv University

The Astrology Sugya (Shabbat 156)
Initial Observations by Dovi Seldowitz

"In this Talmudic segment (sugya) it seems that astrological thinking seems to be assumed by the Sages as legitimate wisdom. At the same time, there is no single perspective which is considered as the ideal interpretation. For example, Rabbi Yehoshua ben Levi is of the opinion that the weekday on which one is born determines their destiny. Rabbi Chanina argues that it is not the day of the week but the hour of the day. And to complicate matters, some Talmudic Sages claim that the Jewish People are not bound the rules of astrology (although they would agree that these rules apply to non-Jews)."

"If it is the case that the Sages who insist that there is no astrological impact on the Jewish People proclaim this view on account of their denial of astrological efficacy, perhaps this can explain how other Sages seem to uphold astrology. These supporters of astrology may be aligned with the state of Judaism from the era of the Hebrew Prophets and therefore maintain that these spiritual forces are still at play."

Disclaimer:
I am not condoning the practice of astrology as it is commonly known today. However, we must understand what the Magi believed in order to realize what motivated them to risk their lives and undertake the long journey to Bethlehem.

Daniel 6:1-4 *It pleased Darius to set over the kingdom one hundred and twenty satraps, to be over the whole kingdom; 2 and over these, three governors, of whom **Daniel** was one, that the satraps might give account to them, so that the king would suffer no loss. 3 Then this **Daniel** distinguished himself above the governors and satraps, because an excellent spirit was in him; and the king gave thought to setting him over the whole realm. 4 So the governors and satraps sought to find some charge against **Daniel** concerning the kingdom; but they could find no charge or* *fault, because he was faithful; nor was there any error or fault found in him.*

At the time of when the Persian king Darius conquered Babylon, Daniel could not have been less than 80 years old, he was still retained by the new regime having a position of high responsibility. He was so highly respected; he was made one of the three presidents who superintended the respective governors of Persia's 120 provinces. (Dan. 6:1-2)

Based upon the Babylonian Talmud and its teachings in astrology and view of Jupiter and Saturn, it is a strong indication that the Magi (Matt 2:1-9) were not pagan and more than likely Jews and descendants of those who were captive in Babylon.

According to Herodotus (1:101), Magi existed in Persia in the sixth century BC, they were one of six tribes and formed the hereditary priestly clan among the Medes who performed religious ceremonies and interpreted signs and harbingers. Persia (now Iran) conquered neighboring Mesopotamia Babylon (now Iraq) and from the fourth century BC forward Magi were increasingly associated with astronomy and astrology. Babylon was the world center of astronomy and astrology at that time, and Magi were important members of the Babylonian royal court as seen in the book of Daniel. From the time of the Exile onward, Babylon contained a strong Jewish colony. It is probable that the knowledge of the Jewish prophecies of a Savior-King, the Messiah, was well-known both to the Babylonians and to the Magi.

The Magi are the prime example of how we are to view the stars, as part of the great astronomical clock, the timepiece God ordained, which He reveals in Genesis 1:14, Psalm 19:1-4, and Psalm 104:19. How do we know they are the definitive example? By their actions and the results of those actions. When the Magi saw the star, they rejoiced; they followed it, and when they found the child, they worshiped Him (Matthew 2:10–11).

The Magi observed, understood astronomy and astrology, and knew exactly whom to worship!

Chapter 13

Tradition

From Origen to Ignatius, the Church Father's non-biblical writings contain dramatizations which increased over time as to the brightness of the star, fueled only by imagination, that grossly exaggerated the star and have influenced the Christmas story, nativity scenes, Christmas Carols and Christmas Cards. However, the Gospel of Matthew makes it clear that "His star" was so insignificant that Herod had to be told about it, and Jerusalem was surprised, having seen nothing out of the ordinary.

The early Church Fathers with their deficient understanding of astronomy, had their own ideas as to the time of year Jesus was born. The writer Epiphanius (AD 315-403), wrote that a religious group from Asia Minor called the Alogi, believed June 20 or May 21, AD 9, as the birth date of Jesus. Epiphanius thought, that this was the day of conception, and that Christ was born on the eighth day before the ides of January, thirteen days after the winter solstice and the beginning of the increase of the light and day. He was convinced he had the birth date as January 6, but could not figure out the conception date, and leaned towards the 20th of June. Unfortunately, that is only a seven-month pregnancy, so he wavered between June and March 20$^{th.}$ The dates the Church Fathers calculated, were consistently sticking to old tradition with Christ being conceived in spring and born in midwinter.

We must be careful not to cling too heavily to the traditional portrayal of what we have been told was the Star of Bethlehem or Nativity scene. Starting with and adhering to Scripture, we move on to see if there is historical, archaeological and astronomical evidence. Notice I did not say evidence to support a particular tradition or my own version, but "evidence." As Mondo Gonzales the Director of the Psalm 19 Project and cohost of the Prophecy Watchers says, *"Let the evidence speak for itself."*

"Don't let Tradition be the Enemy of Evidence"
— *AC Katz*

Chapter 14

Evidence

Scripture and Evidence, not speculation determines the dates to investigate "His star" astronomically.

The Dates

Based on Scripture, historical writings and astronomy we can say with substantial certainty, the date of Herod's death must be March – April, 4 BC. The early part of 4 BC as the date for Herod's death is the consensus of many scholars. Luke speaks of Mary and Joseph traveling to Bethlehem for the census of Caesar Augustus. Caesar Augustus called for a census of **the whole Roman Empire in 8 BC**, recorded in the *Res Gestae* in the very words of Caesar Augustus: "A second time, in the consulship of Gaius Censorinus and Gaius Asinius (8 BC), I again performed the lustrum". (Lustrum: a sacrifice offered at the end of a census.)

A census of that magnitude of the whole Roman world, would have taken at least 2 years to complete. From scriptural, historical, and archaeological evidence, we can be sure that Jesus was born during the time period between 8 BC and 6 BC. With 7 BC as the most likely date we should look into the heavens to see what may have occurred astronomically from the "second census" of Caesar Augustus in 8 BC to 6 BC, during the time Herod was still alive and when Jesus would have been under the age of 2 years.

Chapter 15

His Star

In Matthew 2:1-11, there is nothing to indicate that the star was extraordinary. Herod did not know about it. Nothing unnatural appeared in the sky, and he had to ask the Magi when "His star" first appeared. It must have been subtle and followed normal celestial mechanics in order to go unnoticed.

The Magi said, "We saw His star." It couldn't have been a supernova. A supernova is unpredictable, and therefore could not have been considered "His star." Nova's are not wandering, they remain stationary in relation to the fixed stars, so this possibility is unsatisfactory and must be rejected.
A comet could not be "His star", a comet does not remain in the sky long enough and does not fit the description by the Magi of "His star" in Matthew.

The Bethlehem Star could not be a supernatural unforeseen event to be "His star".

The Magi were astronomers and astrologers, "His star" was a series of celestial events involving stars, planets (wandering stars) and conjunctions. They would have been particularly interested in a conjunction involving the "king planet" Jupiter and Saturn *(Shabbtai),* the planet in charge of the Jews. The Magi had been tracking the star since they had seen "His star in the east." They watched and noted the stars and their relationship to each other and their trajectory and positions within the constellations. Each constellation had a meaning. The Magi already knew what "His Star" was, narrowing it down to only what they could view without a telescope, the visible planets, which they called wandering stars.

Babylonian astronomers were limited by what they could see with the naked eye, the telescope was not invented until the

Renaissance. Although we are not entirely sure who to give the credit to, the first person to apply for a patent for a telescope was Dutch eyeglass maker Hans Lippershey in 1608. What this meant in planetary terminology, is at that time astronomers could only observe the celestial movements of Mercury, Venus, Mars, Saturn and Jupiter, but not Uranus, Neptune and Pluto. The Babylonian astronomers were advanced in their observations particularly regarding any phenomena involving these five planets, and they are recorded in many Babylonian astronomical almanacs that survive to this present day. They recorded and accurately predicted in advance lunar and solar eclipses, solstices and equinoxes, the first and last dates stars would be visible in the night sky, the positions of the planets in relation to zodiacal signs, conjunctions (when celestial objects appear closest to each other), and oppositions (planets appearing on the opposite side of Earth from the Sun).

Matthew 2:2 *"Where is the One who is born King of the Jews? For we saw his star in the east (anatole) and have come to worship him."* The Magi said they saw "his star" in the East, in original Greek this is "ton astera en te anatole" "anatole" translated literally, in its rising.

Note the difference in Greek spellings.

Matthew 2:1 N-GFP
GRK: μάγοι ἀπὸ **ἀνατολῶν** παρεγένοντο εἰς
KJV: wise men **from** *the east* to Jerusalem,

As compared to the others a different word is used.
Matthew 2:2 N-DFS
GRK: ἐν τῇ **ἀνατολῇ** καὶ ἤλθομεν
KJV: His star **in** *the east,* and are come

Matthew 2:9 N-DFW
GRK: ἐν τῇ **ἀνατολῇ** προῆγεν αὐτούς
KJV: they had seen **in** *the east,* went before

When they saw the star they rejoiced exceedingly with great joy!

Matthew 2:10

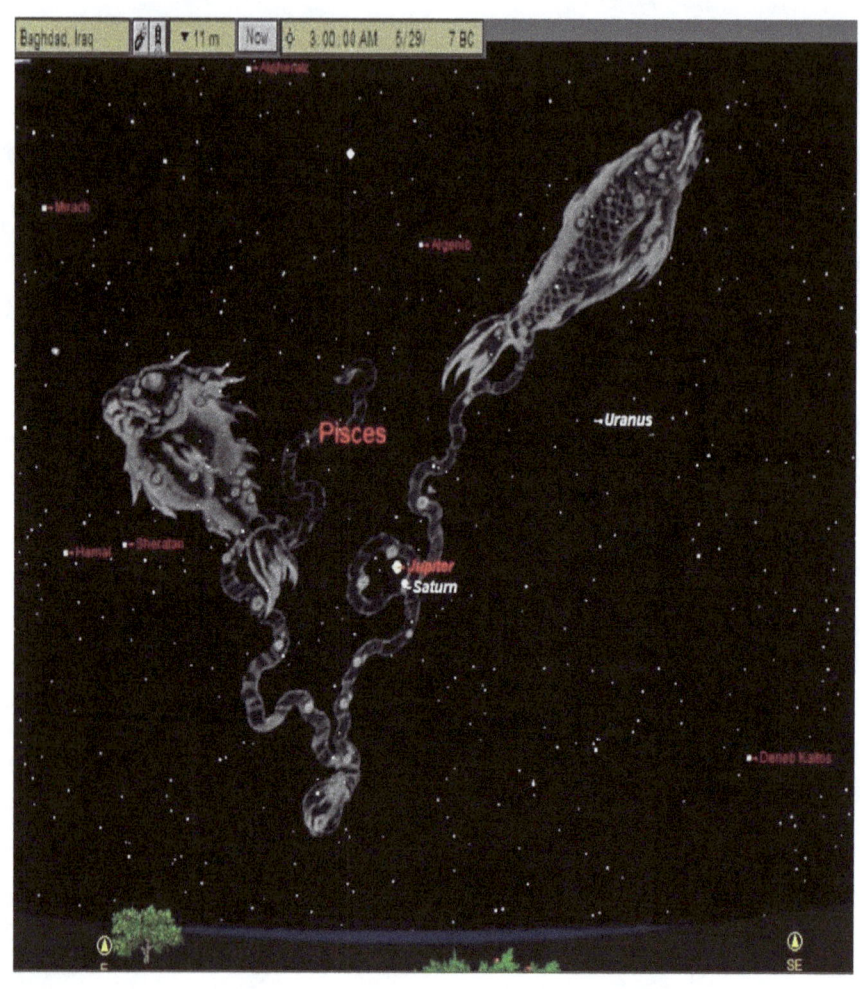

East   West
First Conjunction
May 29, 7 BC
Viewed from Babylon

Chapter 16

The Candidate

Using the dates calculated from the evidence, my husband, an engineer, an expert in orbital mechanics, and an amateur astronomer for many years, used Starry Night, an astronomical simulation program, to observe the skies during the period from Caesar Augustus's census in 8 BC to the death of Herod in March–April 4 BC. He traced the triple conjunction of Saturn and Jupiter in 7 BC. Using this technology, which was unavailable in previous years, we could confirm more precise dates and **times.**

The Sequence
During late 8 BC and early 7 BC the planets Saturn and Jupiter are coming together at a rate of nearly 3.5 degrees per month.

First Conjunction
Date May 29, 7 BC
From Babylon, in the morning Saturn and Jupiter rise in the east. Looking directly due east 3:00 a.m., Saturn and Jupiter are in conjunction at 0.95 degrees and rose due east in Pisces. It was a moonless night, making the sky completely dark; therefore, the conjunction would have been seen and stood out as brilliant in the night sky. A new moon is dark, because it is in between the Earth and the Sun, the backside of the moon is lit by the Sun, this would have been considered a moonless night. The light of the Sun is reflected on Saturn and Jupiter. The Magi would have seen this as the beginning, they also had calculated—evidenced by the Babylonian tablets— what would take place next that same year in the night sky and what it meant, prompting them to plan their journey.

After the first conjunction, later on July 27, 7 BC Saturn and Jupiter separated and their distance increased to 3.0 degrees apart.

Second Conjunction

September 27, 7 BC viewed from Babylon, Saturn and Jupiter are in conjunction at 1.0 degree **in Pisces again**, and is first observed in the sky in the east at 6:50 p.m. while the Sun sets in the west. Saturn and Jupiter are heading east to west, then they set later in the west at 4:51 a.m. Viewed from Babylon, the star had traveled from east to west, "westward leading" and set over Israel. The conjunction of Saturn and Jupiter become visible in Jerusalem at sunset 6:27 p.m. Then both planets set at 5:27 a.m., they were very bright in the dark night sky, dark because in Jerusalem it was again a new moon. Jupiter the "king"star, and immediately underneath is Saturn meaning the "Jews". Jupiter is coronating Saturn, "the King of the Jews!"

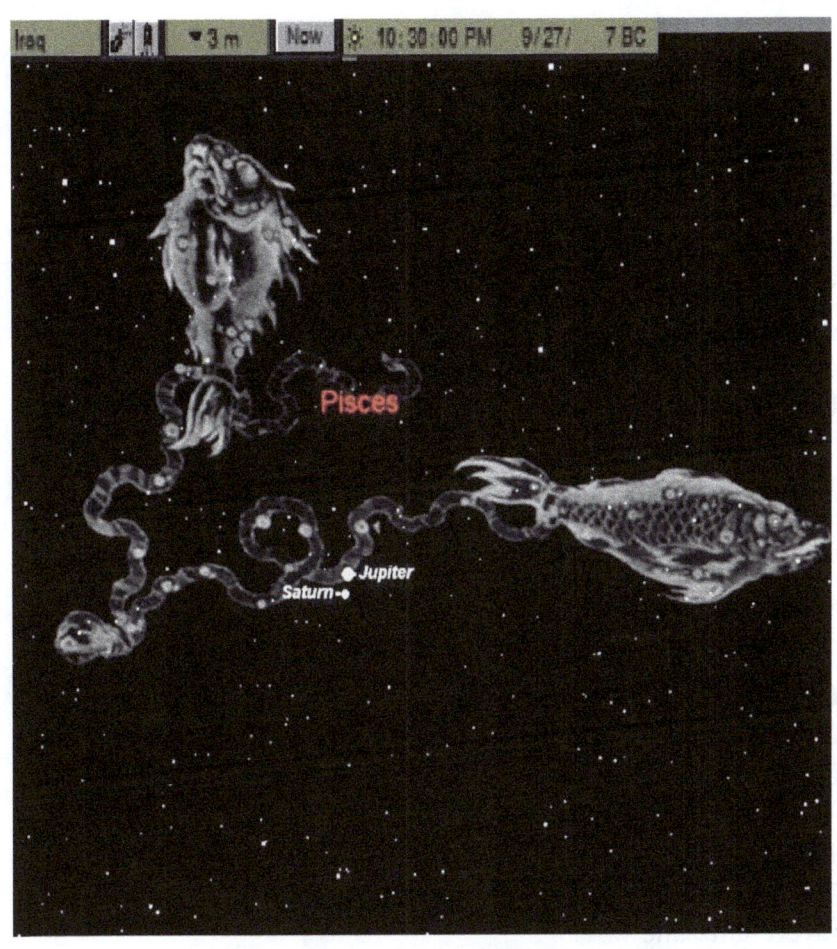

Second Conjunction
September 27, 7 BC

It rose in the east, traversed across the night sky
and set in the west.

While confirming the date of the second conjunction from Jerusalem, we viewed the first sighting of a sliver of a new moon at 6:00 p.m. in the night sky, this was also Rosh Hashana. Was it an announcement, a birth announcement, perhaps? This may have been what prompted the Magi to leave on their trip westward.

Feast of Trumpets in 7 BC occurred on the first of Tishri, with the first sliver of the new moon appearing on September 27, 7 BC, according to the lunar cycles and the Hebrew calendar. Jupiter, the king star, was coronating Saturn, the star of the Jews, as king. The "King of the Jews" was **consecrated** in Pisces on Rosh Hashanah!

Here's what makes Rosh Hashanah a "celebration" in its modern form, even with its solemn roots in the "Day of Blasting":

A Sacred Alarm: The shofar blast served as a divine alarm clock, interrupting daily life to call the people to attention before God.

The Anniversary of Creation: The rabbis identified this day as the anniversary of the creation of Adam and Eve, making it the "birthday" of humanity, and thus the logical "New Year" for all people.

A Day of Coronation: The shofar blasts were also interpreted as fanfare for God, the King, as the Jewish people accept His sovereignty anew.

Rosh Hashanah is deeply joyful, but its joy is intertwined with awe, introspection.

It is a celebration, but not a party. It's a celebration of the most profound and solemn kind. The shofar blasts are not just an alarm clock; they are also **fanfare**.

Rabbinical tradition frames Rosh Hashanah as the day humanity crowns God as the sovereign ruler of the universe.

This is a joyous, celebratory act. The mood is one of awe and majesty, like attending the coronation of a beloved and just king.

First Sighting of the New Moon
from Jerusalem
September 27, 7 BC

The same night shortly after sunset
the new moon is seen setting in the west.

October - November: After the second conjunction and forward into November, Saturn and Jupiter barely separated from 1.1 to 1.2 degrees.

December 1st, 7 BC, this date has been chosen by some as the choice for the third conjunction. The geocentric view, as viewed from Earth by the Magi, Saturn and Jupiter were at 1.1 degrees, at sunset the Moon is half illuminated, which washes out the stars in the sky while still in the southeast.

By December 5th, Saturn and Jupiter were separated by 1.2 degrees, and the Moon was ¾ full at sunset and very bright. It was not a dark night, and this would not have produced a Zodiacal light (downward beam of light) which will be explained in the next chapter.

Third Conjunction
December 20th, 7 BC viewed from Jerusalem, Saturn and Jupiter rises at 12:22 p.m. mid-day but are not yet viewable. The Magi leave Herod and head to Bethlehem. **Twilight begins at 6 p.m. And behold, the star they had seen in the east went on before them! At 6:30 p.m. Saturn and Jupiter are in conjunction** at 1.2 degrees in the moonless dark night (new moon) and are visible in the sky due south at 45 degrees up to Zenith appearing over Bethlehem. Jupiter the king star is crowning Saturn the star of the Jews. The Sun had just set, and the Zodiacal light appears! With the Zodiacal light lasting for approximately 2 ½ hours after sunset. This is also the night before the winter solstice. "This is the only and most suitable condition for the zodiacal light to shine down on Bethlehem, no other date or hour works, that is how significant it is!" —GK

An interesting note is this was also during Hanukkah the "Festival of Lights".

Third Conjunction
December 20th, 7 BC
From Jerusalem

The Difference Between the ongoing Conjunction and Dates of December 5th and December 20th

December 5th

Saturn and Jupiter are closest together, they appear at twilight at 6 p.m. but in the southeastern sky. December 5th the Moon is ¾ full and too bright, Jupiter is southeast which puts it at an angle farther away from the Sun when the Sun sets. The conditions were not suitable for a zodiacal light to occur.

On December 20th when Saturn and Jupiter appear at 6 p.m. twilight they are overhead due south over Bethlehem as viewed by the Magi and at a shallow angle to the sunset (also a moonless dark night) making the conditions most optimal for a zodiacal light to form directly over Bethlehem. This would occur at no other time, all of this taking place while in the middle of Pisces. Saturn and Jupiter remain in conjunction at 1.1 to 1.2 degrees, they **stopped** separating while in retrograde over Bethlehem.

Separation

December 21, 7 BC viewed from Jerusalem, Saturn and Jupiter had risen in the east during the day unable to be viewed because of the sunlight, Saturn and Jupiter became visible at 7 p.m. Saturn and Jupiter during the dark skies of a new moon and the Sun setting in the west, began to emerge at their conjunction of 1.3 degrees in the center of Pisces. It appears due south 45 degrees up to Zenith, then sets over Bethlehem at 11:45 p.m.

By December 23, Saturn and Jupiter began to separate and were at 1.4 degrees, the separation would be recognizable to the Magi. They continue to move apart quickly in early 6 BC at the same rate they came together in early 7 BC.

December 25, 7 BC they are now further apart at 1.6 degrees.

Finally, Jupiter leaves Pisces and crosses into the constellation Aries on March 8, 6 BC.
As Jupiter just leaves Pisces both Jupiter and Saturn are in close proximity of the Sun and can't be seen.
(verified in Starry Night software)

The first and second conjunctions began in the morning in the east leading to the west, the third conjunction in December is an evening event in the west over Bethlehem.

In the Minds of the Magi

From September to December Saturn and Jupiter are two stars on a journey, each day getting closer and closer signifying something was going to happen. In the minds of the Magi, when they become closest together it marks an event. The Magi as astrologers watched the wandering stars, unlike the other stars, the wandering stars (planets) were always moving. At that time, they would have watched them intently, because these wandering stars moving across the dark sky, were of exceptional interest to them. Saturn and Jupiter coming near to each other would be spectacular, compared to the other stars of the night which are stationary. These planets are moving in the celestial sphere unlike normal fixed stars. Their movement made them different; they were considered messengers pouring forth speech and revealing knowledge of coming appointed times. (Psalm 19:1-4 Genesis 1:14) More importantly, the whole event was being played out in front of the constellation Pisces, to the Magi this represents the house of the Jews!

Magnitude of Brightness

Matthew does not describe the brightness of the star. During the second conjunction on September 7 BC, Saturn and Jupiter due to being in opposition, **Saturn was 38 times brighter** and **Jupiter 500 times brighter** than the surrounding stars in Pisces. The brightness of Saturn varies depending on its angle and viewability of its rings. The angle depends on the orbital year of

the Earth. Calculating backward from the ring viewing angle, it was found that Saturn was at -7 degrees in 7 BC, making Saturn viewable from its north pole downward. The brightness of "His star" would have been spectacular and **a very rare event!**
As Saturn and Jupiter were coming together, the intensity of their brightness was significantly increasing due to being in opposition at that time and the position of Saturn's rings. What a message and call to action!
Magnitude information source:
Dr. David W Hughes,
Department of Physics and Astronomy University of Sheffield
The Star of Bethlehem.

There is importance in understanding the astrological traditions of the Magi and ancient astronomers, Jupiter was the planet associated with kings, while Saturn was the protector of the Jews and the planet which reigned over their religion.
(Babylonian Talmud)
You might ask "But there were two planets, how is this one star?" If it had only been one planet, it wouldn't have meant as much as when two are in conjunction, the two most important stars to the Magi. Jupiter says "king" Saturn says, "of the Jews" and Pisces says "Israel". It has been said that "His star" in Matthew was simply **Saturn**, the star of the Jews and symbol of their Messiah. As mentioned previously there is the well-known link in the Babylonian Talmud between Saturday (Shabbat in Hebrew) and **Saturn** (Shabbetai in Hebrew), affirming **Saturn** as the planet in charge of the Jews.

In our attempt to avoid "astrology" one could say giving the planets and stars names or meanings is a sin, let's remember they are the creation of God, He calls them good, and He calls them by name, (Psalm 147:4).

Isaiah 40:26 *Lift up your eyes and look to the heavens: Who created all these? He who brings out the starry host one by one and calls forth each of them by name. Because of his great power and mighty strength, not one of them is missing.* (NIV)

Job 9:7-8 *Who speaks to the sun so it does not rise, and seals up the stars; He alone spreads out the heavens, and treads on the waves of the sea; He makes the Bear, Orion and Pleiades, and the constellations of the south.* (TLV)

God answered Job
Job 38:31-32 *Can you bind the chains of the Pleiades or loosen the belt of Orion? Do you bring out the constellations in their season or guide the Bear with her cubs?*

Job 26:13 *By His spirit he hath garnished the heavens; His hand hath formed the crooked serpent.*
(KJV)

Isaiah 45:11-12 *Thus says the Lord, the Holy One of Israel, and its Maker: How dare you question Me in the work of my hands? It is I who made the earth and created man upon it. It was My hands that stretched out the heavens, and I ordained all their host.* (NKJV)

The "Triple Conjunction" Phenomenon

What made the 7 BC event special was that it was a "triple conjunction." Due to the apparent retrograde (backward) motion of both planets as seen from Earth, Jupiter and Saturn passed by each other three times in a short period.

May 29, 7 BC	Pisces
September 27, 7 BC	Pisces
December 20, 7 BC	Pisces

Frequency Summary

Single Jupiter and Saturn Conjunction	Every 20 years
Triple Conjunction	Every 140 years (approx.)
Triple Conjunction in Pisces	Once every 900 years.

This specific astronomical sign the Magi witnessed, a triple conjunction of Jupiter and Saturn in the constellation Pisces was a once-in-a-millenium event.

This immense rarity perfectly explains why it would have been interpreted by knowledgeable Babylonian astrologers as a sign of unparalleled importance, signaling the birth of an eternal king in Judea (a meaning associated with Jupiter, Saturn, and the sign of Pisces in their tradition). It was not just another planetary meeting; **it was a once-in-an-era celestial announcement!**

Chapter 17

Zodiacal Light

The very word *Zodiac* can cause people to panic, since it is so often associated with astrology. Ancient civilizations referred to the pathway of the constellations as the "pathway of animals," because to them, the constellations marked the path along which the Sun, Moon, and planets traveled. The word zodiac is derived from the Greek phrase meaning "animal circle" but some attribute it as originally a Babylonian concept created in 5 BC. The Zodiac refers to the specific twelve constellations along the ecliptic that the Sun passes through once per year. They are constellations, but not all constellations are zodiac constellations, many constellations are located outside the ecliptic plane. Babylonian astronomers divided the band of sky, the ecliptic plane where the Sun and planets move, into 12 segments which occupy 30 degrees of celestial longitude. The twelve segments indicate the twelve months of the year and were given names known as "the zodiac signs".

"Zodiacal light" are words to describe a natural phenomenon caused by the Sun and interplanetary dust. The dust grains, whatever their origin, spread out in space from the Sun in a flat disk of space and narrow pathway called the ecliptic. This flat space around the Sun is the plane of our solar system and is inhabited by Mercury, Venus, Earth, Mars, Jupiter and Saturn. (Uranus and Neptune but not visible with naked eye) The Solar System planets lie very close to the ecliptic plane, their orbits tilt from the ecliptic plane by a few degrees at the most. Mercury has the largest tilt at only 7 degrees and other planets range from the least Uranus at 0.8 to 3.2 degrees. The Sun seen from Earth travels this same pathway with the Moon at a 5.1-degree tilt crosses the ecliptic plane, these two points are called nodes (the crossing at the nodes are where an eclipse occurs) as they make their journey across our sky.

What does Zodiacal light look like?
Zodiacal light is a diffuse glow in the night sky that appears as a triangular band of light, like a cone extending upward from the horizon typically along the ecliptic, the apparent path of the Sun through the zodiac constellations. This light is most easily observed just after sunset or just before sunrise only during certain times of the year. (an example below)

EarthSky community photos: Christoph Stopka in Westcliffe, Colorado March 2022

The phenomenon is caused by sunlight scattering off tiny interplanetary dust particles that orbit the Sun, **mainly between the Earth and Jupiter.** These dust particles are thought to be remnants of comet tails and asteroid collisions. The dust is concentrated and viewed in the plane of the solar system, the light typically follows the ecliptic, making it appear in the zodiacal band.

But in the case of a comet, the zodiacal light can be in any orientation due to the comets orbit around the Sun and its the trail of debris it leaves behind. Zodiacal light can be mistaken for the first light of dawn or the last light of twilight, but unlike twilight, it is produced by interplanetary dust rather than atmospheric scattering. It is most visible in dark skies away from city lights, where light pollution is minimal.

Bethlehem at the time of the birth of Jesus had almost zero light pollution, no electricity with the faint glow of oil lamps within small homes, unlike cities today. It would have been very dark skies as the Sun set, during a new moon (moon free sky). In low latitude regions, such as Israel, it appears in the western sky after the end of the evening twilight (sunset).

The narrow cone of light stands up above the horizon and reaches **halfway to the zenith.** It shares the apparent diurnal motion of the sky and during the night sinks below the western horizon until approximately two hours after its first appearance at the end of twilight, only the upper portion remains visible, which is the more narrow point of light.

With the Earth rotating at 15 degrees per hour, December 20th 7 BC the king star Jupiter with Saturn beneath it, was at 45 degrees **halfway to Zenith**, it will take approximately two and a half hours to set. But it gets better!

With Bethlehem located at a higher elevation, the horizon would have appeared higher to the Magi ascending towards the light as they entered Bethlehem on the northeastern side. They would have seen the light narrowly concentrated on the spot where the child was.

Matthew 2:9 Original Greek *"And having heard the king, they went away, and behold the star which they had seen in the east, went before them, until having arrived it **stood** over the place where the child was."* During the extreme of retrograde.

December 20th 7 BC Details

Saturn and Jupiter are in conjunction, at this time, they appeared due south of Jerusalem at sunset, pointing the way (as verified by Starry Night). As the Magi traveled south on the Hebron Rd, their journey taking them south-southwest to Bethlehem. During this 2 ½ hour distance Jupiter and Saturn with Zodiacal light began to set from south to southwest as verified in Starry night.

The arc of the ecliptic changes every month and is only on a true east-west during the summer and fall equinoxes two days out of the year. Due to the December time period, it was winter solstice, the longest night and shortest day, when the ecliptic is low in the horizon. Therefore, it sets in the southwest rather than true west, matching the testimony of the Magi. It was pointing to Bethlehem as they traveled along the road. At this time, the planets are in extreme retrograde over each other, stopping over (stood, stood still in reference to the background stars). Jupiter stopped over Bethlehem and crowned Saturn on that day, and proceeded in their final separation (as verified by Starry Night and the Babylonian Tablets of 7 BC).

The Magi were the only ones (in Jerusalem) who **recognized, understood** and **properly interpreted it.** Because they were astronomers and *magoi* (wise men) from Babylon with the understanding of what Saturn, Jupiter, and Pisces meant.

Some people have gone so far as to speculate the star was an angel flying overhead that led them to Bethlehem. Angels don't typically rise in the east; they did not see an angel in "anatole" in its rising. The Magi said they saw His star, *astera*.

It does not say in the Scriptures, that the Magi said it was directly overhead or that it led them. "It went on before them" which is technically correct as to where the star (conjunction) was positioned ahead of them from their position on the road to Bethlehem. As they were entering Bethlehem, the star had stopped, was in retrograde, and had begun to set. The Magi understood retrograde, it is clearly recorded in the Babylonian Tablets of 7 BC and hundreds of years before.

Professor William F. Albright (American Biblical archaeologist and Middle Eastern scholar) estimated that Bethlehem had at most a population of about 200 to 300 people when Jesus was born. At about 6 to 10 people per household, that represents approximately 25- 35 households spread out over two hilly plateaus.

No matter what way you look at it, retrograde and/or setting, the star stopped over the location of a little house among 25 to 35 houses scattered over the hills of Bethlehem. Coming up the road and entrance to Bethlehem, from the viewpoint of the Magi the star set and due to the homes being spread out, it would appear over the very area of the house and courtyard. They were entering a village that sat at an elevation of 2550 feet, making the star look as if it stood above the house.

There is no way to prove with 100% certainty that this was the Star of Bethlehem without a living witness. The Magi were the knowledgeable witnesses of that time, the best astronomers of their day—and *Starry Night* is a tool to validate their testimony.

There was **no** Zodiacal Light during the second conjunction, the suggested birth time of Jesus. The second conjunction was not in position for a Zodiacal light to occur.

Zodiacal Light was possible only during the third conjunction December 20th, and during the time involving the arrival of the Magi and would shine down from the Star to the place where the child was.

Viewing a Zodiacal light over Bethlehem in the dark moonless night sky would have been brilliant for the 2 ½ hour period! Somewhat difficult to view today due to light pollution and other criteria. Looking at some of the photos of Zodiacal light that have been photographed in modernity, we cannot get a picture of the full brilliance of the light that occurred over Bethlehem that holy night long ago.

Different comets have different paths. We know of comets and their previous trajectories because some were visible with the naked eye and recorded by ancient astronomers. Due to advancement in technology, we can calculate their previous and future trajectories and track comets that have varying trajectories, including small comets that are not visible, although they are still within our inner solar system.

Zodiacal light is a product of the Sun's light illuminating the dust particles, they are typically left over from the tails of previous comets, the dust particles are left in their orbital trail, and they do not disappear over time.

Is it possible that a comet traversed the sky in previous years leaving a trail of dust particles in its orbital path that **would have caused a Zodiacal light on that night December 20th, 7 BC to be even more luminous and brilliant?**

Chapter 18

The Helper

The comet GIACOBINI ZINNER (21 P) 21P is a mid-sized comet whose orbit features a relatively short period, low inclination, and is controlled by Jupiter's gravitational effects. 21P/ Giacobini-Zinner orbits the Sun every 2,390 days (6.54 years), coming as close as 1.01 AU and reaching as far as 5.99 AU from the Sun, its orbit is highly elliptical. The comet traversed through Jupiter and Earth in 12 BC this comet was not visible to the Babylonians or anyone else, it could not be seen with the naked eye, it was a fast medium sized comet, it just had one job to do, and that job was to leave the debris trail in its orbital pathway in between the Earth and Jupiter. It wasn't discovered until December 20, 1900, with a telescope.

The question is, was this comet Giacobini-Zinner in our inner solar system or way out beyond? Its path needed to be within our inner solar system to have left dust between Jupiter and Earth.

Let's take a look at its characteristics. Comet 21P/ Giacobini-Zinner is a medium sized comet with a diameter of 1.24 miles (2 kilometers). This comet takes about 6.6 years to orbit the Sun once. It is what you would call a fast comet, certainly not a comet that would hesitate over the town of Bethlehem, this by itself would not be the Bethlehem Star, but that is not what we are looking for. We want to know, was it within the inner solar system and did it leave dust on its orbital path between Jupiter and Earth?

In 1985 (Sept. 11) a re-purposed mission named ICE (International Commentary Explorer, formally International Sun-Earth Explorer-3 ([ISEE-3]), was sent to gather data from this comet. Each time that the comet Giacobini-Zinner returns to the inner solar system its nucleus sprays ice and rock into space.

In 12 BC it came closest to the Sun in its orbit, and very close to Earth right between Earth and Jupiter. This would have left a massive particle dust trail on its orbital path as though it spread icing on a cake. And would have profoundly contributed to the brilliance and luminosity of the Zodiacal light that shone down from Jupiter through Saturn above Bethlehem, visible only during the hours after the Magi left Jerusalem, as the Sun began to set at 6:30 p.m. December 20th, 7 BC.

This comet returned in 5 BC, but Jupiter and Saturn were not in conjunction, they were also no longer in the constellation Pisces, more importantly the comet wouldn't have been visible without a telescope, no one could see it. It is evident today with the annual meteor shower that the debris field still exists. It is clear this comet had only one job to do, frost the cake. The comet 21P/GZ did its job!

What Causes a Comet's Tail?
Solar wind is not the type of wind we often think of on Earth. Solar wind consists of mainly protons and electrons (not to confuse with photons that have no mass) Asteroids and comets are affected by solar winds, these protons and electrons act like bullets, as they hit a comet, they strip off dust and ice creating the comet's tail. It depends on the Sun's cycle, or any CME's (Coronal Mass Ejections) that take place during a comet's orbit as it gets closer to the Sun. With comet 21P/GZ, as it traverses by the Sun and in between the Earth and Jupiter it leaves behind a trail of ice and dust particles. Electromagnetic radiation from the Sun creates heat on the side of the comet or asteroid facing the Sun, which can cause it to roll and change its trajectory. As could be the case for Apophis in its future flyby in 2029. There are too many variables that can affect its trajectory. They will not know more about its trajectory until it is closer. One thing I can predict about Apophis is that it is unpredictable One thing I can predict about Apophis is, it is unpredictable— at least presently.

[Jupiter-family Comets | Space Reference](#)

The many Jupiter-family comets and their orbits:
Scientific data shows Zodiacal light reflects from comet dust trails left in their orbits. The dust from a comet's orbit responsible for a Zodiacal light to occur is not limited to the ecliptic plane.

"Jupiter-family Comets are objects whose orbit features a relatively short period, low inclination, and is controlled by Jupiter's gravitational effects. There are 786 Jupiter-family Comets in this database out of 1,302,506 total, accounting for 0.1% of objects."

Jupiter-family Comets | Space Reference

The orbit of Giacobini-Zinner

The data shows that this comet 21P/GZ, comes close to Earth and spews the dust and debris for the Sun to illuminate during zodiacal light. This comet is a good candidate because it is a medium sized comet, not visible to the naked eye, it is fast moving, which means it comes around more often replenishing the dust trail, and it passes through the ecliptic plane relative to Earth and Jupiter. Some examples of recent close approaches in the 20th and 21st century highlight its Earth Jupiter alignment. Hence why it is called a Jupiter family comet.

The comet experienced seven close approaches to Earth and two close approaches to Jupiter during the 20th century. (From the orbital work of Kazuo Kinoshita)

- 0.88 AU from Earth on 1900 December 15 (contributed to comet's first discovery)
- 0.51 AU from Earth on 1913 November 14 (contributed to comet's second discovery)
- 0.26 AU from Earth on 1946 September 20
- 0.93 AU from Jupiter on 1958 January 19
  - decreased perihelion distance from 0.99 AU to 0.94 AU
  - decreased orbital period from 6.56 to 6.42 years
- 0.35 AU from Earth on 1959 November 8
- 0.58 AU from Jupiter on 1969 September 23
  - increased perihelion distance from 0.93 AU to 0.99 AU
  - increased orbital period from 6.41 to 6.52 years
- 0.93 AU from Earth on 1972 July 24
- 0.47 AU from Earth on 1985 September 6
- 0.85 AU from Earth on 1998 November 27
- 0.39 AU from Earth on 2018 September 11

[21P/Giacobini-Zinner (cometography.com)](cometography.com)

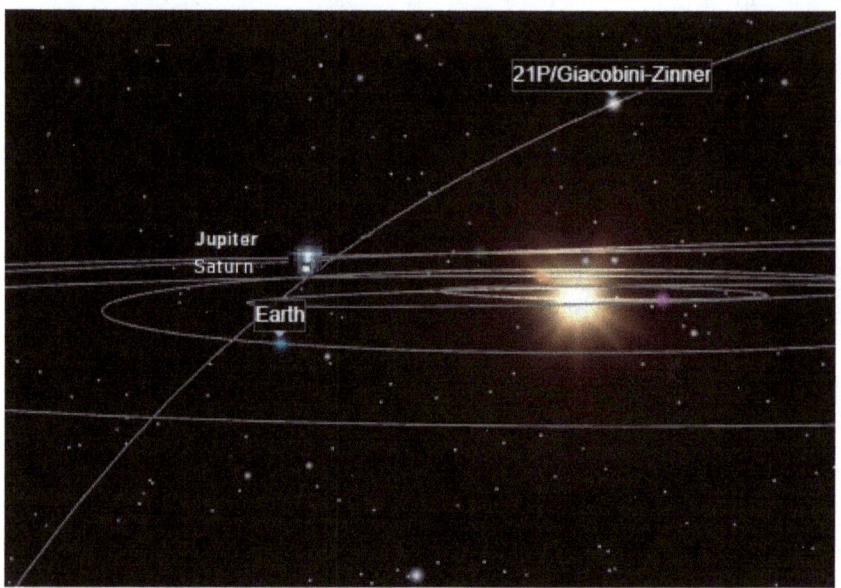

December 20, 7 BC

"Zodiacal light normally reflects from dust within the ecliptic plane. Scientists and data have shown Zodiacal light reflecting from comet dust trails left in their orbits. It's a sea of dust, and we are looking through it as the Sun sets. G-Z 21/P is a fast-moving comet that orbits every six years, continuously replenishing its dust trail. On December 20th, 7 BC, Earth was aligned with Jupiter and Saturn at the thickest part of the dust lane, due to the closest proximity of the comet's orbit around the Sun. This makes it extremely rare and brilliantly visible in the sky."
—*GK, amateur astronomer*

Israel is located in the northern hemisphere, north of the equator. It is situated at latitude 31.0460. The latitude is the position relative to the equator specifying the north-south position. December 21st, is the **winter solstice** along with the moonless sky on December 20th, this would have occurred during the darkest night of the year.

From the road to Bethlehem, as the Sun is setting a cone of light appears to extend upward from the western horizon shortly after sunset, it appears to slant upon the dust trail of the comet (through the path of the planets) the slant produces an inverse cone of light shining from the star, the wandering star Jupiter (representing the king) through Saturn (the Jews) directly underneath it and pointed downward to the place where the child was.

Have all the Christmas cards with the Christmas Star been wrong? Not entirely.
We must remember that the star is not mentioned in the Gospel of Luke 2:8-12, at the time of Jesus' birth, when the shepherds were told "And this *will be* the sign to you: You will find a baby wrapped in swaddling cloths, lying in a manger." However, in Matthew, "His star," shining brilliantly, had a purpose: it was to direct the Magi to the place where the child was —which is exactly what took place. With this evidence, we find that during the third conjunction on December 20, 7 BC, the brightness of "His star" and the alignment in the sky could have produced Zodiacal light shining downward on Bethlehem, confirming for the Magi the location of the *King of the Jews*. And that the Bible, the Gospel of Matthew, and the Bethlehem Star is scientifically, astronomically, geographically, and magnificently accurate!

Chapter 19

Bethlehem

Luke 2:4-7 (NKJV) *Joseph also went up from Galilee, out of the city of Nazareth, into Judea, to the city of David, which is called Bethlehem, because he was of the house and lineage of David, to be registered with Mary, his betrothed wife, who was with child. So it was, that while they were there, the days were completed for her to be delivered. And she brought forth her firstborn Son, and wrapped Him in swaddling cloths, and laid Him in a manger, because there was no room for them in the inn.* (Greek word used, Kataluma =guest room).

September 27, 7 BC: A Birth Announcement?
The night of the second conjunction.

    Joseph and Mary had traveled to Bethlehem, a little village of approximately 200 people 300 at most, at that time.It may have consisted of barely 3 blocks of houses (in modern comparison), spread out with some grouped in clusters with approximately 10 people to a household, or two homes with extended families being connected by a courtyard. When Joseph arrived, he would go to his family home or neighborhood. The *Kataluma* (the guest room) was full and the house busy with in-laws, aunts, uncles or cousins, no room in the *Kataluma*.
    It is highly likely a relative of Joseph, a sister or aunt would have said "take your wife to the lower level of the house where the animals are usually kept" or "across our courtyard to the small shelter or cave where the sheep are kept, they are not there, but out in the field, it is empty, clean, and there is fresh straw/grass stored to use for bedding. There is a stone manger to line with fleece or dry grass for the baby to lie in." As they settled in the evening, Mary delivered her first born son, the conjunction of

Jupiter and Saturn became visible in Jerusalem **at sunset 6:27 p.m**. As the Sun set, it could clearly be seen, Jupiter had crowned Saturn, the star of the *King of the Jews*, while in Pisces. The first sliver of the new moon appeard in the dark night sky to the west — it's Rosh Hashana!

The Moon

The Moon serves not only as a sign (*ot*) but also a *moed* – a Hebrew word best translated as "appointed time". This is the time God himself determined for an appointment with mankind.

The constellation **"Pisces,"** which of course means, "Fish" is the Roman name that is most familiar. But the Hebrew name for this sign is, **"Dagim,"** which means, "Fish (plural)."

Pisces/ Dagim is represented by the 10th letter of the Hebrew alphabet Yod.

Yod is a representation of an outstretched hand implying **receiving** something or recovering something. To help, hand, arm of the Lord, work, heal, power, authority.

Yod

As in the Exodus, a previous moed was the giving of the Torah— the Law— to Moses on Mount Sinai. Now, Jesus the Messiah, the very fulfillment of the Law, was also given.

Romans 10:4 (NKJV) *For Christ is the end of the law for righteousness to everyone that believes.* God's hand is the ultimate authority in the land, relating to the recovery of the lost sheep of Israel. Matthew 15:24 *But He answered them and said "I am not sent but to the lost sheep of the house of Israel."*

One Rabbi explains, *until the Shacharit Amidah, with a few notable exceptions, the morning prayer is the same as every Shabbat and holiday.* The difference is that during the year, we feel more distant, **while on Rosh Hashanah, we understand that our King is very near to us.**

"In Bethlehem of Judea," the Torah scholars replied, when Herod inquired where the Messiah was to be born. The prophet Micah issued a Messianic precursor, his encouragement proclaimed God's coming Kingdom and the coming Messiah, Jesus. Micah specifically announced the birthplace of the Messiah as Bethlehem.

Micah 5:2 *"But you, Bethlehem Ephrathah, though you are little among the thousands of Judah, yet out of you shall come forth to me the one to be ruler in Israel, whose goings forth are from of old, From everlasting."* (NKJV)

Ephrath refers to Bethlehem, the first mention of Ephrath occurs in Genesis 35:19, referring to the place where the Biblical matriarch Rachel died giving birth to Benjamin, she was buried on the road from Bethel. Evidence that she died on the way there, is indicated by the ancient "Rachel's Tomb" at the city's entrance. It is clearly not Bethlehem of Galilee, as some have suggested.

Rachel's Tomb

An old postcard of Bethlehem

**Bethlehem-from-Star-St-c 1880-George-Al-Alma-Collection
Photo by-Felix-Bonfils-Source**

Entering Bethlehem on Christmas Day

Old photo in Bethlehem

Bethlehem 1860 Old Road going up into Bethlehem

The previous photos of Bethlehem are from the 1800's forward, during the time of Jesus it was far less populated.

Bethlehem during the Birth of Jesus

Luke 2:7 records, "there was no room for them in the inn (kataluma= guest room)." The only other New Testament reference to *kataluma* was the upper chamber "guest room" of a Jerusalem house where the Last Supper was held (Luke 22:11 – along with its parallel in Mark 14:14; (NIV, NASB, NKJV). There is no reason why Luke's use of *kataluma* in 22:11 ("guest room") on the last night of Jesus' life should be different from his use of the same term, in 2:7 *kataluma,* on the first night of His life. In fact, there may even be some symbolism with these "guest rooms" serving as bookends to His life and ministry. There was no room for Him in a *kataluma* on the first night of His life and He presided over a meal (probably a Passover meal) in a *kataluma* on the last night of His life.

Luke's statement "no room for them in the inn" suggests the "guest room" of the house where Mary and Joseph were now staying was already full. The NIV's 2011 revision acknowledged this fact, changing "inn" to its true meaning "guest room" (2:7). It may well be that older relatives were already settled there and/or Mary preferred not climbing to an upper floor or appreciated a bit more privacy in the domestic stable below, cave under the house or across their small courtyard.

Away in a Manger, But Not in a Barn: An Archaeological Look at the Nativity.(biblearchaeology.org) Greg Byers MA The Life & Ministry of the Lord Jesus Christ & the Apostles 26-99 AD

Kataluma the original word Luke used is a Greek word it does not translate as Inn, there were no Innkeepers mentioned in Luke when Mary and Joseph entered Bethlehem for the birth of Jesus. Nowhere does it mention a stable or manger in Matthew by **the time the Magi arrived.**

Inns and Innkeepers The good Samaritan. Luke 10:34 *So he went to him and bandaged his wounds, pouring on oil and wine; and he set him on his own animal, brought him to an inn and took care of him.* Luke used the Greek word pandocheion for public inn. Luke 10:35 uses the Greek word *pandochei* for "innkeeper". This is different than the word he used in **Luke 2:7** *kataluma,* which would be a guest room in a private home.

Homes commonly had four rooms, there was a main entrance typically opening into a courtyard (usually unroofed). The longer rooms are divided from each other by wooden post on stone bases. There may have been low curtains to divide the rooms. Usually, animals were kept in the room below and would be taken out during the day. As Scripture said, the night of the Jesus birth the sheep were out with the shepherds at night. Therefore, a lower room would have been clean with fresh bedding and a manger. The upper floor room would be the upper "guest room" (*kataluma*) in (Luke 2:7).

"While the biblical text makes no mention of a cave, neither does it mention a stable; Luke simply records that baby Jesus was laid in a manger (Luke 2:7, 12, 16). People have assumed the unmentioned stable because of the presence of a manger. However, numerous permanent stone-carved or plastered stone-built mangers have been discovered on the ground floor of houses from biblical times. People in ancient Israel would sometimes keep young, vulnerable, or special animals safe inside the home at night.

The archaeological and textual evidence suggests that Mary and Joseph were not stuffed in a barn out behind the local Motel 6 in Bethlehem, but rather occupied the manger room on the ground floor of a relative's house because the upper room was already occupied. There need not be a discrepancy between a cave and a manger room in a house, as some domestic structures were built against a cave, which would have housed the manger room.

Ancient Stone Manger

The biblical text, written tradition, and the archaeological evidence all affirm that Bethlehem of Judea was indeed occupied during the first century AD. The gospels of Matthew and Luke explicitly state that Jesus was born in Bethlehem of Judea, fulfilling the ancient prophecies that the Christ would be born in Bethlehem Ephrathah. The tradition of the birth cave is mentioned in writings from the second, third, and fourth centuries, culminating with the construction of the Church of the Nativity. In summarizing the evidence, archaeologist John McRay stated, "The tradition that Jesus' birth took place in Bethlehem [of Judea] is long and solid. There appears little reason to doubt its essential trustworthiness."
Source Bible Archaeological Report
O Little Town of Bethlehem
December 10, 2021
Bryan Windle

"And this will be the sign to you: You will find a baby wrapped in swaddling cloths, lying in a manger."

Luke 2:12

Context in Luke 2:12: **The Night of Jesus' birth.**
The angel tells **the shepherds:** "And this will be a sign for you: you will find a baby (brephos) wrapped in swaddling cloths and **lying in a manger.**"

In Greek, ***brephos*** specifically refers to a **newborn infant**, a very young baby.

This is a powerful sign because finding a *newborn baby* lying in an animal's feeding trough is unexpected and highlights the humble circumstances of Jesus's birth.

Comparing Scriptures

Luke 2:12 The night of Jesus' birth.

Matthew 2:11 The night the Magi visit.

Context in Matthew 2:11: **The visit of the Magi**

Matthew 2:11 "And when they had come in to the **house,** they saw the young child *(paidion)* **with Mary His mother**, and fell down and worshiped Him."

This is different from the word ***brephos*** used in Luke 2:12 for the newborn Jesus to the shepherds.

In Greek, ***paidion*** when the Magi visit, refers to a very young child of a few months old to a toddler under two years old.

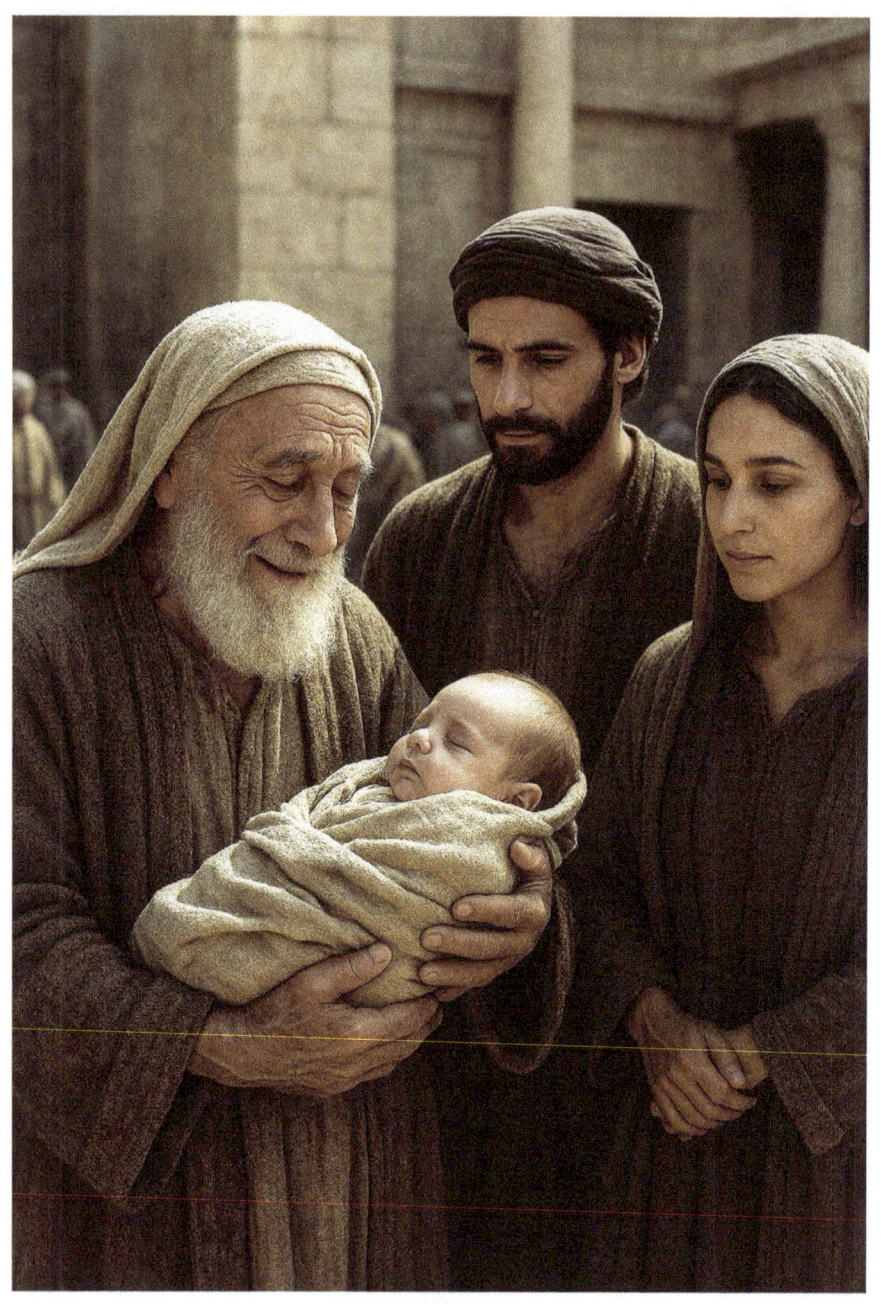

His Name

"When the eight days before His circumcision had passed, He was named Jesus, the name the angel had given Him before He was conceived. And when the time of purification according to the Law of Moses was complete, His parents brought Him to Jerusalem to present Him to the Lord,(as it is written in the Law of the Lord: "Every firstborn male shall be consecrated to the Lord"), and to offer the sacrifice specified in the Law of the Lord: "A pair of turtledoves or two young pigeons."

"Now there was a man in Jerusalem named Simeon, who was righteous and devout. He was waiting for the consolation of Israel, and the Holy Spirit was upon him. Led by the Spirit, he went into the temple courts. And when the parents brought in the child Jesus to do for Him what was customary under the Law, Simeon took Him in his arms and blessed God, saying: "Sovereign Lord, as You have promised, You now dismiss Your servant in peace. For my eyes have seen Your salvation,which You have prepared in the sight of all people, a light for revelation to the Gentiles, and for glory to Your people Israel." Luke 2:21-32

Forty days after Jesus was born, fulfilling the purification requirement of Leviticus 12, Mary and Joseph travel to Jerusalem's temple to present Him before God. The trip is only 6 miles (9.6 kilometers) long. His parents make an offering at the temple of two young birds. It is during their visit that a priest named **Simeon** prophesied about Jesus' mission in life and blessed his parents. If the **Magi** had already visited and presented their gifts, under Jewish law Mary and Joseph would have been obligated to give more than a pair of turtledoves.

"Then Herod called the Magi secretly and learned from them the exact time the star had appeared. He sent them to Bethlehem and said, "Go and search carefully for the child. As soon as you find him, report to me, so that I too may go and worship him."

Matthew 2:7-8

Chapter 20

The Star and Visit of the Magi

December 20, 7 BC, a phenomenon occurred in the evening sky over Bethlehem!

Herod, after hearing about the prophecy in Micah 5:2 that Messiah was to be born in Bethlehem, sent the Magi to Bethlehem, (Matthew 2:7,8). As the Magi left in the direction of Bethlehem, while the Sun was beginning to set at 6:00 pm, the star began to emerge in the twilight, in the profound darkness of the moonless Middle Eastern desert sky, the brilliance of the planets Saturn and Jupiter appeared over Bethlehem on the horizon. Shortly after sunset just as the stars began to appear in the night, a cone of light appeared, called a Zodiacal light, as the darkness continued in the desert night, the light became more prominent, it produced an inverse cone of light which appeared to be shining from the star, similar to a flashlight downward. With the intercession of this rare Zodiacal light, it would have appeared to point persistently to a certain point on Earth. While traveling on the road the Magi would view the star above as it appeared (at that time) due south above Bethlehem at 45 degrees up "halfway to Zenith". The cone of light appeared to point downward and for at least two hours continued to shine getting brighter and brighter as the skies darkened around Bethlehem.

As the planets continued to set and the cone of light moved with the diurnal motion of the sky, it began to appear south-southwest (it was the winter solstice). The cone of light became lower and continued to become narrower and more direct as the Magi continued on the path. As they ascended the path and arrived, the star would appear to stop, and the cone of light would have continued to point, directing the zodiacal light from the star above downward to the place where the child was. We can calculate an approximate time the Magi arrived since the road to Bethlehem was about 2 1/2 hours by foot. They left Jerusalem before the star appeared, which was just before sunset that day at 6 p.m. After the Magi "went their way" while they were on the road the star became clearly visible and appeared over Bethlehem at 6:30 p.m.

As they were arriving in Bethlehem sometime between 6:30 and 9:00 p.m., with Bethlehem at a higher elevation, the horizon would have risen closer to the narrow point of the conical-shaped Zodiacal light as they approached from the road, making it appear as though the star stood directly over the location where the child was.

After listening to the king they went their way. And behold, the star they had seen in the east went on before them, until it stopped over the place where the child was, when they saw the star, they rejoiced with exceedingly great joy. Matthew 2:9

As explained in a previous chapter, at that time with the estimated population of Bethlehem at about 200 to 300 people and with 25-35 homes spread out or in groups, the star would have easily been seen stopping and setting above the place where the child was (Scripture does not say over a stable or manger).

After Jesus was born, they were dwelling in *"the house."* Bethlehem, at that time being a very tiny village, meant that everyone knew where everyone lived and who they were. Joseph had brought Mary to his hometown. Since Bethlehem is elevated, people would have seen a group of interesting strangers entering the village in the evening. A relative, or at least a friend, would have asked them whom they were seeking. After the shepherds had already visited a few months earlier, everyone would have known who the Magi were looking for. As the Magi traveled the road south to Bethlehem, the star appeared over Bethlehem the entire way.

The Road to Bethlehem

The Magi would have traveled a route from Jerusalem to Bethlehem that followed well-established paths used by travelers and caravans during that time. Bethlehem is about 6 miles (9.6 kilometers) south of Jerusalem, so the journey would not have been long. The route would have taken them along a road that passed through the Judean hills. Historically, travelers would have likely followed the main road south from Jerusalem, which would lead them directly to Bethlehem. Considering the short distance between the two cities, the journey would have been relatively uncomplicated.

The ancient route the Magi likely took from Jerusalem to Bethlehem corresponds to modern-day roads that follow similar paths. The most direct modern route is along Hebron Road, also known as Route 60. This road begins in the southern part of Jerusalem and heads towards Bethlehem, following a path that is believed to be close to the one taken by travelers in ancient times. Hebron Road is a major artery connecting Jerusalem with the southern parts of the West Bank, including Bethlehem, and it roughly traces the route that has been used for centuries. Today, it's a well-traveled highway, passing through parts of East Jerusalem and continuing through the Judean hills into Bethlehem. The ancient road would have been less developed but would have followed a similar geographical trajectory due to the natural contours of the land.

After the Magi left King Herod, they left Jerusalem and headed south southwest to Bethlehem.

Three Men on the road to Bethlehem 1800's

South-southwest on Hebron Road

Bethlehem is located at an elevation of about 775 meters (2,543ft) above sea level, 30 meters (98) higher than nearby Jerusalem. Bethlehem is situated on the Judean Mountains. Its horizon as viewed from the road would have been elevated and appeared closer to the top of the conical zodiacal light even more so just as the Magi arrived in Bethlehem from the road coming from Jerusalem.

    The dwellings in the little village of Bethlehem were on the crest of the western ridge of the elevation, the eastern ascent on the Jerusalem side was lightly inhabited and contained the grotto at the very site where the Church of the Nativity now stands.
His star and the zodiacal light appear to stop and stand over the grotto at the Church of the Nativity.
It doesn't get better than this!

(Streets in yellow are main roads)

The Magi came from Jerusalem on Hebron Rd., then onto the main old road which is now Manger St. Following the topography to King Davids Wells (for water), continuing on the road then ascending on Beit Sahour St. placed the Magi looking southwest, the star would appear to be setting over the Church of the Nativity or that area.

Is the Church of the Nativity the place of Jesus' birth? I have not been one to rely on tradition. Nevertheless, the star stopped (was in retrograde) and would have appeared above while entering Bethlehem and its light shone down on the place where the child was, it just so happens that the Church of the Nativity is in that location. The birth could have taken place there, or within close proximity to that area. The point was not to prove the Church of the Nativity as Jesus' birthplace. In fact, I have been the biggest skeptic; And yet—it is not only possible, it is likely, every piece of data aligns. The goal was not to find the exact spot of His birthplace, but to acknowledge that God is in control of nature: past, present, and future. This was an appointed time, a planned event, programmed into the heavens during creation.

By observing we find inexpressible joy realizing He has the power to control His designs. It would be frightening if God did not have a plan or care for us, but our worries are quieted regarding present day events and noticeable disorders in nature, when we know that all things are under the direction and control of God, and His great designs are made to accomplish His plans.

Isaiah 46:9-10 God says *"I am God, and there is no other; I am God, and there is none like me. I make known the end from the beginning, from ancient times, what is still to come. I say, 'My purpose will stand, and I will do all that I please.'"* (NIV)

How long would the Magi's journey take?
Scripture gives us an idea. They first saw the star in conjunction when they were in the east on May 29, 7 BC and arrived in Bethlehem on December 20, 7 BC. They could have started their journey any time after viewing "His star" in its first conjunction in May, but due to sweltering heat in the Middle East would have started their journey no earlier than late September.

The second conjunction in September likely would have prompted them to begin their journey when travel would have been better.

We learn from Ezra 7:9 Ezra began his *aliyah* journey back to Jerusalem from Babylon the first day of the first month Nisan (mid-March). His large entourage, which likely included the elderly and children, took four months to travel from Babylon to Jerusalem

The popular thought of there having been three wise men, is based on the three gifts and has no bearing on the actual size of the group that traveled with the Magi. Being that they were all adults on a mission, it is doubtful there were elderly or children included, and their trip would have taken less time than Ezra's to complete. We have seen in some studies, there would have been at least twelve people in the group, for the care and watering of the animals and the needed protection from thieves along the way. We know it was a very noticeable entourage by the stir they caused when they arrived in Jerusalem.

This timing fits well for them, having seen the second conjunction September 27, 7 BC to embark on their journey during the time of year when the weather made it possible for travel. And more importantly, Saturn and Jupiter in conjunction, was emerging in the evening and clearly heading from the east and was westward leading, setting west over Jerusalem. This actually ties in with the fact that Mary and Joseph were in Bethlehem for at least 40 days, until the end of Mary's 'purification'. (Luke 2:22). It seems that the couple stayed on in Bethlehem for a while, surrounded, as they would have been, by Joseph's family. According to the timing of the first conjunction on May 7 BC and second conjunction on September 7 BC, Jesus could have been anywhere from 4 months old to 9 months old when the Magi arrived **December 20, 7 BC.**

The Family living in Bethlehem
and
The Arrival of the Magi

When the Magi arrived in Bethlehem, Jesus' family was living in a house, not a stable (Matthew 2:11). The wise men from the East, guided by "His star," came into the house to worship the "King of the Jews," Jesus, who was with His mother, Mary.

"And when they had come into the house,
they saw the young child with Mary His mother,
and fell down and worshiped Him.
And when they had opened their treasures,
they presented gifts to Him:
gold, frankincense, and myrrh."

Matthew 2:11

"And when they had come in to the **house**, they saw the young child with Mary His mother, and fell down and worshiped Him."

Matthew 2:11

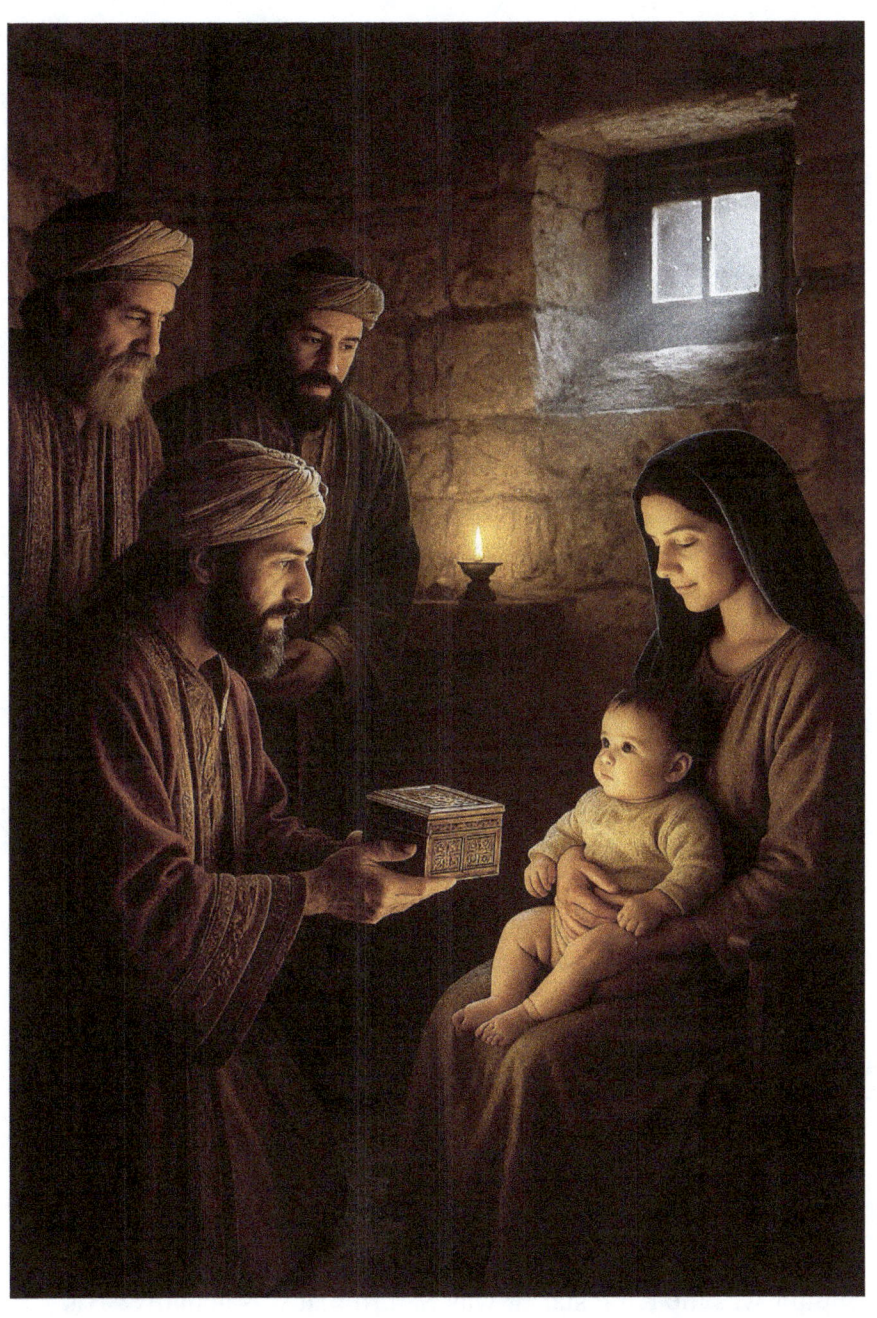

Why did Herod choose two years old and under?
So often it is assumed that the child Jesus was two years old when Joseph and Mary fled to Egypt, because of the threat from Herod to kill all the male children two years old and under. Nowhere in Scripture does it say the actual age of Jesus at that time. He could have been a couple of months old to two years old. Matthew 2:7-8 *Then Herod called the Magi secretly and found out from them the **exact** time the star had appeared. He sent them to Bethlehem and said, "Go and search carefully for the child. As soon as you find him, report to me, so that I too may go and worship him."* (NIV)

After the Magi had left.
Matthew 2:13-15 *Now when they had departed, behold, an angel of the Lord appeared to Joseph in a dream, saying, "Arise, take the young Child and His mother, flee to Egypt, and stay there until I bring you word; for Herod will seek the young Child to destroy Him." When he arose, he took the young Child and His mother by night and departed for Egypt, and was there until the death of Herod, that it might be fulfilled which was spoken by the Lord through the prophet, saying, "Out of Egypt I called My Son."* (NKJV)

Matthew 2:12 And (the Magi) being warned of God in a dream that they should not return to Herod, they departed into their own country another way.

Matthew *2:16 When Herod realized that he had been outwitted by the Magi, he was furious, and he gave orders to kill all the boys in Bethlehem and its vicinity who were two years old and under,* **in accordance with the time he had learned from the Magi.** (NIV)

As was previously stated, the understanding of the Magi was "Saturn" (*Shabatai* "or Shabbat") was the star of the Jews and Jupiter was the King star; it was Saturn that was actually "His star".

Matthew 2:1-2 (NKJV) *Now after Jesus was born in Bethlehem of Judea in the days of Herod the king, behold, wise men from the East (Greek anatolon) came to Jerusalem, saying, "Where is He who has been born King of the Jews? For we have seen **his star** in the East (Greek anatole) and have come to worship Him."*
The Greek text says the Star was *"en anatole,"* meaning they saw His star rising in the east.

The date when the Magi would have first seen "his star" Saturn, in its rising in the east was July 21st of 8 BC, rising at 10:23 p.m. Saturn rises in the east, Jupiter is heading towards Saturn, no conjunction yet, but Saturn "His Star" is now in its rising in Pisces. (This does not mean Jesus born in 8 BC) Matthew 2:7 *Then Herod secretly called the magi and determined from them the **exact time the star had appeared.*** The Magi informed Herod of the "exact time" they saw His star in its rising, this is the only sighting they mention to Herod. At this time, the Magi and Herod did not know the age of the child. We do know, that based on their information, Herod determined to have all the boys two years and under killed. Hearing of the timing of Saturn in its rising July 21st, 8 BC, during Herod's interrogation of the Magi during the third conjunction of December 20, 7 BC, in Herod's mind, that would have made the child about one and a half years old (17 months, the amount if time His star remained).

The Magi did not return to Herod; after going to Bethlehem they went home another way. Herod would have realized he was fooled possibly a month or two afterward January, February, or March. Since the entourage of the Magi was likely at least 12 or more people, the Magi could have stayed in Bethlehem for more than a few days to feed and water animals and prepare for the return trip. Nevertheless, whenever the Magi left, Mary and Joseph left afterward for Egypt, with Jesus being from 5 to 10 months old. Based on the "exact time" Herod was told "His star" was seen by the Magi the previous year on July 21st, 8 BC. From this time, it would make Herod understand the child to now be **19 months old.** He would have called for the death of the male toddlers under two years old in January- March the beginning of

6 BC. The exact timing the star appeared, gave Herod the information to kill the male children under two years old, two years would cover all those children born since the Magi first "saw His star in its rising" in the east.

Two years later (after the visit from the Magi Dec 7 BC) and Herod's realization he was fooled, it was spring of 4 BC. Herod had died and Archeleus became Ethnarch, as history and Scripture says, soon after his father Herod died.

Both Origen and Eusebius state that Jesus and his family were in Egypt for two years and they returned in the **first year** of the reign of Archelaus. In Herod's will, Archelaus, one of Herod's sons, would start his reign after his death. History shows Archelaus started his reign in 4 BC which also confirms Herod died between the lunar eclipse of March 13 and Passover April 11, in 4 BC. According to all of the facts, the 1 BC eclipse is three years too late to reconcile with Josephus in Antiquities, Origen, Eusebius and the reign of Herod's successors in 4 BC.

The first year of the reign of Archelaus was from April 4 BC forward that year. Jesus and his family had left for Egypt after the Magi left Bethlehem, which was likely January- March 6 BC while Herod was still alive. Returning after Pentecost of 4 BC, after Herod was dead, the uprising against Sabinus and Archelaus became Ethnarch. This would have been in the first year of the reign of Archelaus, having spent about two years in Egypt as Origen and Eusebius state. Thus, the triple conjunction and birth of Jesus in 7 BC is consistent chronologically with both Herod's massacre of the infants and the two-year stay in Egypt, as is written in Scripture and historical writings. Matthew 2:16-18
"Then was fulfilled what was spoken by Jeremiah the prophet, saying: 'A voice was heard in Ramah, lamentation, weeping, and great mourning, Rachel weeping for her children, refusing to be comforted, because they are no more'" (NKJV)
(Pentecost uprising Timeline: Flavius Josephus, Antiquities of the Jews, Book XVII, Chapter 7, Section 1)

An Occultation of the Moon over Jupiter in 6 BC

While investigating the orbital path of Jupiter through the month of March in 6 BC, we observed the occultation of the Moon over Jupiter. Although it does not have the characteristics of the Bethlehem Star, it was invisible and could not be a star the Magi "saw", nor was it seen over Bethlehem.

With this, I **only suggest** another possibility; instead, it may have represented the flight of the Holy family into Egypt. The occultation on March 20, 6 BC occurred after the Magi would have arrived December 20, 7 BC during the third conjunction of Saturn and Jupiter in Pisces stopping over Bethlehem. At this time the Magi had returned another way, Herod realized he had been fooled, Joseph was warned in a dream to flee to Egypt. The occultation of March 20, 6 BC the Moon was very low in the sky, as soon as Jupiter came above the horizon at the rising of the Sun, the dark Moon, like a curtain proceeded to move towards Jupiter (king star) which could not be seen in the sunlight. The Moon progressed to cover and hide Jupiter so their light could not be seen, they both could not be viewed. The occultation began at 4:00 pm and remained covered and unviewable as the Sun set at 7:00 pm. Jupiter **could not be seen,** it was traversing/traveling in secret behind the dark moon. The light of the "King of the Jews" (Jupiter), was now darkened as viewed from Bethlehem, his light was no longer there. As we read in Scripture, Joseph had taken Jesus and his mother Mary and fled "during the night" after sunset. This also fits with timeline as Origen and Eusebius said the Holy family was in Egypt for two years. Two years later in 4 BC Archelaus was reigning in Judea.

Matthew 2:19-23
After Herod died, an angel of the Lord appeared in a dream to Joseph in Egypt and said, "Get up, take the child and his mother and go to the land of Israel, for those who were trying to take the child's life are dead." So he got up, took the child and his mother and went to the land of Israel.
But when he heard that Archelaus was reigning in Judea in place of his father Herod, he was afraid to go there. Having been warned in a dream, he withdrew to the district of Galilee, and he went and lived in a town called Nazareth.

In regard to an occultation of the Moon, ancient astronomers believed the star went inside the Moon. "Finally, when a star (wandering star = planet) may disappear behind the Moon; the Akkadian phrase is, it "entered" the Moon. Sometimes the point of entrance or exit is specified by expressions like "one third of

the disk to the north". When at the time of observation, a normal star or planet is so close to the Moon that an occultation can be expected (but will not be observable because the Moon will set before it takes place), the star is said to be *ana libbišu kunnu* "set towards its (the Moon's) inside".
Astronomical Diaries and Related Texts from Babylonia
By the late Abraham J Sachs,
Completed and edited by Hermannn Hunger

Weather for the Journey
As far back as 1863, Smith's Bible Dictionary, under the heading 'Palestine: the Climate', explained the rarity of snow in southern Palestine, while it conceded its more frequent occurrence in the northern parts of the land. The mean temperature at Jerusalem during December is said to run around 47 to 60 degrees F.
"The rains come chiefly from the S. or S.W. They commence at the end of October or beginning of November and continue with greater or less constancy till the end of February or middle of March, and occasionally, though rarely, to the end of April.
*It is not a heavy continuous rain, so much as a succession of severe showers or storms with intervening periods of fine bright weather, permitting the grain crops to grow and ripen. And although the season is not divided by any entire cessation of rain for a lengthened interval, as some represent, yet there appears to be a **diminution** in the fall **for a few weeks in December and January,** after which it begins again, and continues during February and till the conclusion of the season."*

Taking into consideration the time of year for the Magi to travel, September to December should be considered a possibility, not only was it a possibility but preferable and highly likely. They were not traveling by camel during the winter across the Northern United States, but in the desert. Having lived in the desert myself, the time of year with least outdoor activity is during the summer months. In Arizona most outdoor activities are scheduled for and increase in the fall.

Many people may have heard of the Scottsdale Arabian Horse Show, which can present 2,400 equine entries and runs for approximately 10 days, this is always in February when the outdoor weather is pleasant, and people travel from all over the world, owners, riders and trainers come from the Middle East, Europe, South America and Australia. The horse show season in Arizona starts in October runs through winter and ends first week of May, riding horses in 105-degree weather in May, is more than unpleasant, but is a health risk for all involved including the animals.

According to The World Climate Guide, weather in the Middle East can have extremely hot summers, they are hotter on average than Arizona. It's not uncommon for temperatures to reach 120 degrees or more in those places during July & August. In Iraq, (Babylon) the climate is desert in the center and the south, with mild winters and extremely hot summers, and it's semi-desert in the north, with relatively cold winters. During the long summer months, the *shamal* blows, a moderate northwesterly wind, very hot and dry, which may cause rapid dehydration, and when it's more intense, it can raise dust or sand. In the rest of Iraq, which is occupied by plains or hills (zones 2 and 3), the climate is arid mild in winter and scorchingly hot in summer. Summer goes from June to September, but the months of May and October, are hot as well. From May to October it can already approach 104 deg. Although the climate in Mesopotamia is desert, the landscape can change locally due to the rivers, both where they give rise to wetlands, and where they are exploited to irrigate the fields.
Climates to Travel
World Climate Guide
https://www.climatestotravel.com/climate/iraq

Riding camels or horses on a journey in the Middle East from May to August regardless of the route would have been postponed until fall. Therefore, the thought of a wintry wonderland keeping them from traveling would have not been the case, but instead end of September mid-October forward when the beautiful crisp clear night in the desert was so very perfect for viewing the night sky!

Sadly, the significant increase in light pollution over these past 60 or 70 years, is making certain celestial sights much more difficult to see. The time when the weather is muggy, oppressive and miserable is 95% in the month of August, drops to 60% in September, 30% in October and drops in November down to 0% by December. The daily chance of precipitation in Israel in December is only 15 -20% it peaks end of January into February at 24%.

The weather for Herod wasn't the issue for his travel between Jerusalem and Jericho, but shipping was halted during this time of year due to the weather out at sea and politics in Jerusalem at that time (as previously explained) would have kept Herod in Jerusalem through the break in rain in December.

Chapter 21

What Time of Year Was Jesus Born?
From a Shepherd's Perspective

The Shepherds

Luke 2:8-11 *And there were shepherds living out in the fields nearby, keeping watch over their flocks at night. An angel of the Lord appeared to them, and the glory of the Lord shone around them, and they were terrified. But the angel said to them, "Do not be afraid. I bring you good news that will cause great joy for all the people. Today in the town of David a Savior has been born to you; he is the Messiah, the Lord.* (NIV)

    The fields in Bethlehem in the winter were no place for flocks with pregnant ewes (female sheep) about to lamb. In the Talmud it makes it clear that flocks of sheep did not lie out in the fields in the winter months. According to the Talmud those animals are those which **go out into the open in March and return to the pen at the first rains, about November.** It is highly likely the shepherds mentioned in Luke were watching over flocks intended for temple sacrifices. Special care would have been taken during lambing because the flocks were valuable, even if the lambs were not destined for temple sacrifice, and many were, they are all valuable. The rams for breeding or meat, the ewes for breeding, for milk to make cheese, yogurt etc... and both rams and ewes for wool, for spinning yarn for clothing and various textiles.

    Having been a shepherd myself of Basque descent, my great grandfather "Joseph" an immigrant shepherd, lambing season is quite a busy time of year, as most present-day shepherds understand.

    Lambing season is in late winter and early springtime, this isn't always by choice. The ewes (female sheep) become more fertile in the fall due to the shortening of days and the effect of the sunlight on the pituitary gland of the ewe, and is also a time when the ewe is most fertile and presents a higher chance of multiple births. This is the time when the sheep are out in the fields with rams, the months of August, September, and October are warm enough during the night for the flocks to remain in the field. Although the hills in Bethlehem may not have had as much grazing opportunity, as is usually the case when we have our sheep out during the end of summer or fall, it is probable they were in the area of Migdal Eder in the vicinity of the town of Bethlehem where there is stored grass or grain for feed for supplementation. Or any nearby field, rather than being too far out in the barren hills, making it difficult for the shepherds to leave their flock. *"Let's go to Bethlehem and see this thing which the Lord has made know to us!" So they hurried off and found Mary and Joseph, and the baby lying in a manger.* Luke 2:15,16

During lambing season, which is winter and early springtime, December through April even early May, it is the busiest season **in the pen in Israel,** with dozens to hundreds of lambs born in one flock, it is busy anywhere during lambing season. It would be extremely stressful on the ewes to lamb out in the fields. It is very obvious by the physiology of a pregnant ewe to know a few days before she is about to lamb, experienced shepherds know by the size of her udder. To avoid being too descriptive, what takes place under the tail of a ewe soon to lamb, is not only obvious by sight to a shepherd, but also by smell to potential predators. Having lambing ewes in the field would cause an onslaught of predators, including dogs wild or not to become relentless in their pursuit of fresh lamb and other products delivered during and after birth. Before a ewe is about to lamb, she will separate herself from the flock, this would be dangerous for the ewe and newborn lambs and an impossible task for the shepherd to stay near the lambing ewes to protect or assist them, before, during, and after they lamb.

Speaking from experience, not all ewes are capable of lambing without the assistance of a shepherd, more often than expected, the births can be breech. Not all ewes are good mothers, some will reject their lamb or lambs abandoning them or not allowing them to nurse, the lambs must be able to nurse in order to get the colostrum to provide immunity and necessary nutrients for them to survive and grow. Immediately after birth, the lamb is exposed to a variety of infectious agents present in the environment, the ram, and other ewes and lambs. Without any protection from these infectious organisms, the lamb may become diseased or die. They need to nurse preferably within the first 15 minutes after birth, sometimes the lambs are too weak to stand and need assistance from the shepherd. Lambs must have 10% of their body weight in colostrum within the first 24 hours of birth. That means an average 10 lb lamb must receive one pound of colostrum within the first 24 hours, half of this within the first 4 to 8 hours.

Some ewes for various reasons not only reject their lamb, but even physically harm one or more of them by stomping on them or head butting them. In the winter months with hypothermia a continual threat to the lambs, which are born soaking wet, landing in the dirt and mud during rainy season, would not be a desirable birthing location for sacrificial lambs or any lamb. Lambing season is extremely busy, most ewes have twins, triplets and sometimes quads, for the protection of the life of the lambs and ewes, both being very valuable, lambing occurs in the fold (a pen/shelter) in a controlled situation. Any shepherd worth his weight would not consider allowing such valuable lambs to be born out in the fields at night if at all avoidable. During lambing season in winter and early spring, the flocks would not be out at night. In Luke it says they were "out in the fields" with their flocks by night, when Jesus was born. **Therefore, Jesus was not born in spring during lambing season.**

Because Jesus was called the Lamb of God in John 1:29 people have ascribed his birth to lambing season, it is erroneous to equate the birth of Jesus and the shepherds' nighttime watch described by Luke as during lambing season, it is purely speculation and unrealistic.

By late spring to August, September and October, lambs that were born earlier that same year could be 75 to 100 lbs and out with the flocks with less worry for the shepherds of ewes lambing in various places in the fields. This is the time when shepherds would be out with their flocks in the fields at night. Adult sheep can handle the weather all year round in Israel, the priority and focus is lambing and safety of the lambs.

Possible Location of the Shepherds in the Field

Dr. Alfred Edersheim, who considers it likely that the angel appeared to the shepherds at the traditional site, states: "This Migdal Eder was not the watchtower for the ordinary flocks which pastured on the barren sheep-ground beyond Bethlehem but lay close to the town" (Life and Times of Jesus the Messiah 1:186). Although Edersheim may have been correct on other subjects, there is evidently some embellishment in his claims for Midgal Eder being the birthplace of Jesus.

There is no confirmation for the actual location of Migdal Eder other than it lays somewhere in the vicinity of Bethlehem close to town, this may have been a preferred location for the flocks. As a shepherd you must feed your flock, there is not as much grass available in the hills during late summer and fall, they must feed them through the year, nutrition is extremely important for breeding ewes and pregnant ewes, or they will have still born lambs.

We must stay faithful to Scripture. In the Gospel of Luke the angels said the babe would be found in swaddling cloths and laying in a manger. There is no proof to be found that the swaddling cloths his mother Mary used were from old priestly garments or that they prove Jesus was born in Migdal Eder. When a baby is born it is wet and goopy, some people think when lambs are born, they look like cute dry fluffy stuffed animals, but lambs are also born soaking wet and a mess and need to be wiped down. When our lambs are born, after cleaning them up and wiping them down, we place a blanket on them that wraps underneath and around them, similar to a little jacket you might buy for your dog at PetsMart. Why? Because it is wet and cold, not to protect it from breaking its legs or blemishes as Edersheim claims. Human babies have been swaddled for thousands of years in anything from rags to towels or fancy receiving blankets, this does not indicate the location of Jesus' birth. But "in a manger" indicates something out of the ordinary and many homes in Bethlehem had mangers in the lower level of their homes where animals were kept.

Shepherds out in the fields with their flocks,
with Bethlehem in the background.

Chapter 22

Planetary Conjunctions

In astronomy a conjunction is the apparent coming together of two or more planets in the sky when viewed from Earth. They do not actually come together. There is a large distance between Saturn, Jupiter and Earth, but at certain times they come closer in line as viewed from Earth, which is considered a conjunction. Jupiter takes much longer than the Earth to complete its year. Taking 11.86 Earth years to complete its orbit as it travels around the Sun. It is much slower than the earth's orbit and therefore spends approximately 1 year in each constellation as viewed from the Earth, it takes approximately twelve years for it to return to the same constellation.

Let's take a look at the conclusion of Dr David Hughes Astronomer Dr David Hughes University of Sheffield in an interview on BBC program The Sky at Night – The Real Star of Bethlehem. Dr Hughes had spent his life studying the Star of Bethlehem. Dr Hughes explains the planetary conjunction of which occurred in 7 BC as a triple conjunction, "Conjunctions are not rare but a triple conjunction is, Jupiter and Saturn came together within one degree, which astrologically is a conjunction."

This, in itself, is very rare—to be that close. After the third conjunction on December 20, 7 BC, they moved quickly apart and continued moving apart throughout 6 BC.

The Rarity of the Event:
People may say that Saturn and Jupiter have been in conjunction many times over the past 2,000 years. But that overlooks the many **specific details** that occurred in 7 BC. Understanding the combination of all that took place helps explain—and confirm—what made this particular event so unique and spectacular.

Because the planets move in their orbits at different speeds and are located at varying distances, they sometimes appear to pass one another in the night sky. At other times, they may seem to pause or even move backward, a phenomenon astronomers refer to as **retrograde motion**. This trick is like passing a slower car on the highway, as you get close to the other vehicle, it seems to hold still beside you. Then, as you pull away, it drops backward. The same thing happens as Earth moves around the Sun much faster than the outer planets.

Saturn and Jupiter conjunctions are rare. Saturn and Jupiter are outer planets and are much farther away than the other naked-eye planets Mercury, Mars and Venus. Because an object's increased orbital distance, this means it takes more earth time to complete an orbit. The Earth takes one year to complete one orbit around the Sun. Because of the orbital distance of Jupiter, which takes 11.86 years to orbit around the Sun and Saturn which takes 29.5 years, any conjunction of these two planets will happen only every 20 years. Jupiter is closer to the Sun than Saturn, so it also appears to move faster in our night sky. The triple conjunction of 7 BC was special in that the two planets met three times in succession in the same constellation Pisces. Since 7 BC a triple conjunction of Saturn and Jupiter has been observed only twice, in the years 786 and 1583.

During the triple conjunction of Saturn and Jupiter, Jupiter appears to pass Saturn three times, in a zigzag fashion. This phenomenon, is an illusion caused by the Earth's own movement around the Sun. The Sun, the Earth and the two objects lined up in a particular way. What was most important is which planets were in conjunction and where in the sky it was taking place. The Triple conjunction of 7 BC not only occurred in the same constellation during all three conjunctions, but it was in the constellation of Pisces, which had special meaning and made this a unique and outstanding event for the Magi.

Due to the 26,000-year wobble of the earth's axis (perturbation of the North Pole) the Bethlehem star event is astronomically an extremely rare one-time natural event in the 26,000 astronomical cycles.

The famous Johannes Kepler 1571-1630 a German astronomer, inventor, astrologer and mathematician known for the three laws of planetary motion. In 1584, enrolled at the Protestant seminary at Adelberg, with the goal of becoming a minister. In 1589, after obtaining a scholarship, he matriculated to the Protestant University of Tübingen. Kepler is considered one of the most significant contributing minds of the 17th-century scientific revolution. Kepler calculated backward and he also suggested the triple conjunction of 7 BC as the Star of Bethlehem.

In October of 1595, Johannes Kepler conveyed to Tübingen the news that he had completed his first book, the Mysterium cosmographicum. He wrote to Michael Maestlin, his former professor of mathematics, "I truly desire, that these things are published as quickly as possible for the glory of God, who wants to be known from the Book of Nature. I wanted to be a theologian; for a long, time I was distressed: behold God is now celebrated too in my astronomical work." Unable to devote himself to the Book and Scriptures directly, Kepler had turned his focus to God's other book the Book of Nature which, he believed, also revealed God's providential plan.

Does the Triple Conjunction of 7 BC Fulfill the Magi's description of "His star" in Matthew?

With the unique characteristics that occurred in the triple conjunction of 7 BC it fits very well; we see consistency when we stick with the Biblical narrative. The star led the Magi to Jerusalem not Bethlehem. Imagine the Magi arriving in Jerusalem. If it was daytime with the Sun still washing out the sky when they arrived, they would not have seen "His star". Saturn and Jupiter which they had first seen in its rising in the east, this same star, "His star" would now as the Sun set, appear over Bethlehem low in the sky and settled above Bethlehem.

The Magi came first to Jerusalem to inquire about the child, they did not know His exact location. They understood the star heralded the birth of a new King of the Jews, they evidently knew about the prophecy of the birth of the king possibly Isaiah 7:14 and other Scriptures as evidenced by their actions.

Isaiah 7:14 (KJV) *Therefore the Lord himself shall give you a sign; Behold, a virgin shall conceive, and bear a son, and shall call his name Immanuel.*

Scriptural Sequence

Luke 2:1-2 (NKJV) *And it came to pass in those days that a decree went out from Caesar Augustus that all the world should be registered This census first took place while Quirinius was governing Syria.*

Luke 2: 3-6 (NASB) *And all the people were on their way to register for the census, each to his own city. Now Joseph also went up from Galilee, from the city of Nazareth, to Judea, to the city of David which is called Bethlehem, because he was of the house and family of David, in order to register along with Mary, who was betrothed to him, and was pregnant. And she gave birth to her firstborn son; and she wrapped Him in cloths, and laid Him in a manger, because there was no room for them in the kataluma (*guest room*) While they were there, the time came for her to give birth.*

Matthew 2:1 (NASB) *Now after Jesus was born in Bethlehem of Judea in the days of Herod the king, behold, magi from the East arrived in Jerusalem, saying,*

Matt 2:2 'Where is the one who is born king of the Jews?" For we saw his star in the east *and have come to worship Him."*

They did not say the star led them. They already knew the meaning of the star, a new King of the Jews meant Israel.

The Magi did not come directly to Herod, they were going about in Jerusalem inquiring, evidently creating a stir.

Matt 2:3 *"When Herod the king had heard these things, he was troubled, and all Jerusalem with him.*
Then Herod interrogated the priests and experts.
Matt 2:4-9 *And gathering together all the chief priests and scribes of the people, he inquired of them where the Messiah was to be born.*
They said to him, "In Bethlehem of Judea, for so it has been written by the prophet: "And you, Bethlehem, land of Judah, are by no means least among the rulers of Judah; For out of you shall come a ruler who will shepherd My people Israel."
"Then Herod secretly called the magi and determined from them **the exact time** *the star had appeared."*
And he sent them to Bethlehem and said, "Go and search carefully for the Child; and when you have found Him, report to me, so that I too may come and worship Him."
After listening the king, they went their way. And behold, **the star they had seen in the east,** *went before them until it came to rest over the place where the child was."* It does not say stable or manger.
Matt 2:10 Greek *"having* **seen now the star** *they rejoiced with joy great exceedingly"*
Matt 2:11-15 *And after they came* **into the house,** *they saw the Child with His mother Mary; and they fell down and worshiped Him. Then they opened their treasures and presented to Him gifts of gold, frankincense, and myrrh. Then, being divinely warned in a dream that they should not return to Herod, they departed for their own country another way. Now when they had departed, behold, an angel of the Lord appeared to Joseph in a dream, saying, "Arise, take the young Child and His mother, flee to Egypt, and stay there until I bring you word; for Herod will seek the young Child to destroy Him." When he arose, he took the young Child and His mother by night and departed for Egypt, and was there until the death of Herod, that it might be fulfilled which was spoken by the Lord through the prophet, saying, "Out of Egypt I called My Son."*

Matt 2:16- NKJV *Then Herod, when he saw that he was deceived by the wise men, was exceedingly angry; and he sent forth and put to death all the male children who were in Bethlehem and in all its districts, from two years old and under, according to the time which he had determined from the wise men.*

Matt 2:19-23 *Now when Herod was dead, behold, an angel of the Lord appeared in a dream to Joseph in Egypt, saying, "Arise, take the young Child and His mother, and go to the land of Israel, for those who sought the young Child's life are dead." Then he arose, took the young Child and His mother, and came into the land of Israel. But when he heard that Archelaus was reigning over Judea instead of his father Herod, he was afraid to go there. And being warned by God in a dream, he turned aside into the region of Galilee. and came and settled in a city called Nazareth. This happened so that what was spoken through the prophets would be fulfilled: "He will be called a Nazarene."*

A Review of the Astronomical Corresponding Sequence

The 8 BC census of Caesar Augustus was taking place (would have taken at least a year or two to complete)
Joseph and Mary go to Bethlehem.
Herod was alive.
The Magi saw "his star" in its rising, Saturn rose in Pisces July 21, 8 BC in the east also while the Magi were in the east. No matter which way you look at it, in the east or from the east.
First conjunction May 29, 7 BC (in Pisces) Saturn and Jupiter rising in the east.
Second conjunction September 27, 7 BC (in Pisces) rises in the east and sets over Bethlehem, Rosh Hoshana. (likely Jesus' birth)
This second conjunction in its brilliance would have prompted the Magi on their journey. They came to Jerusalem to ask because the religious leaders were not aware of the astronomical event. It was not a comet—it doesn't last long enough; not a supernova—it

wouldn't be called "His star"; and not an eclipse—they don't last long enough either.

When the Magi arrived in Jerusalem during the third conjunction on December 20^{th,} 7 BC, the star became visible at twilight in front of them at 45 degrees up over Bethlehem, it had stopped and stayed in conjunction and set over Bethlehem, during the most suitable conditions for a Zodiacal Light to occur. Beginning at twilight lasting approximately 2 ½ hours after sunset. The conical light shone and became more intense as it set over the place where the child was (Matthew 2:9).
From the first rising of "His star" in the east was a total of 19 months, Herod's reason for having those 2 years and under killed. After December 20^{th,} 7 BC, Saturn and Jupiter separated quickly and for the last time.
The Magi could have left Bethlehem in January (which was only 11 days away) after letting the animals and their group to rest perhaps 7 days or a couple weeks while also gathering a few supplies.
Afterward February- March(at that time travel was slow), Herod realized he was fooled (Matthew 2:16).
Joseph and Mary were warned in a dream and left for Egypt likely February to April and stayed two years.
Two years later 4 BC Herod dies between the Lunar eclipse of March 13, 4 BC and Passover April 11, 6 BC.
The March 13, 4 BC Lunar eclipse is the correct lunar eclipse since it is recorded Herod died at age 70 between a lunar eclipse and Passover. He was born in 74 BC, which means he died in 4 BC.
Joseph and Mary return after two years in Egypt, Archelaus, Herod's son, was in his first year of reign, so they traveled to Nazareth (Matthew 2:19-23).

Chapter 23

Babylonian Records

Solid Evidence
Is there archaeological evidence of when the triple conjunction occurred?

Babylonian records, ancient writings outside the Bible recording the triple conjunction in 7 BC.
   Excavations in Babylon revealed evidence of the conjunction of 7 BC, four clay tablets were uncovered, an almanac. It demonstrates that from the beginning of the year, Saturn and Jupiter were continuously visible together in Pisces for 11 months. It shows Pisces as the background for both of the planets Saturn and Jupiter as they traveled over time through the sky at night. On these tablets is an accurate recording month by month of the movements and stationary points the rising and stetting of both planets.
An eleven-month conjunction of Saturn and Jupiter to within 1.0 degrees while in Pisces is an extremely rare event occurring only **once every 800 years.**

Ancient writings outside the Bible recording of the triple conjunction in 7 BC.
   Babylonians recorded astronomical events on tablets, between 8-1 BC only 1 comet appeared in 5 BC. Comets were not always seen as the birth of a king, also comets do not last long enough. If it was a comet Herod would have seen a comet and yet he did not see anything out of the ordinary. Meteors are too short lived they do not last long enough to be "His star". Another problem with a supernova, is they would not know when it was going to happen. The Magi may have seen a nova, but it would not have had the necessary astrological significance to connect to the birth of the King of the Jews.

Solid Evidence

According to Chris Walker curator of the British Museum, the skilled Babylonians left crucial historical records, the three tablets date 7 to 6 BC real historical evidence.

"Professional Babylonian Scribes, who were hired by the temple for life, to make astronomical observations watching everything that happened in the sky, day and night they performed the mathematical calculations month by month, year by year to predict various astronomical events in the sky and then tell to what kind of things it might relate whether the events having to do with the king or weather affecting crops." (like an almanac)

These three tablets all date from 7 to 6 BC, this is the record of the triple conjunction of 7 BC.

"The tablets say Saturn and Jupiter were in conjunction in Pisces in the year 7 to 6 BC". We have a written record. "It shows us that there were a group of astronomers working at Babylon who would have been able to predict the event and understand its significance." Chris Walker

We need to go back in time to the ancient science of the Magi to understand how they understood the celestial event.

Babylon was rich in culture with men who were taught how to read the heavens. The Magi had a detailed understanding of the movement of the planets in the sky. "They were skilled astronomers." There was specific meaning to this specific event. It held significance that only the Magi understood. The records of what they understood is in the Babylonian Talmud. "To them this was the rarest event."

Bible Archaeology
Israel Restored After the Babylonian Captivity. truthnet.org

Influence of the Planets According to Talmud 156a (Written during the Babylonian captivity)

One who was born under the influence of Venus will be a rich and promiscuous person. What is the reason for this? Because fire was born during the hour of Venus, he will be subject the fire of the evil inclination.

Venus: A Promiscuous person with evil inclination. Venus was not what the Magi would have considered as the star of Bethlehem.

One who was born under the influence of Jupiter [tzedek] will be a just person [tzadkan]. Rav Naḥman bar Yitzḥak said: And just in this context means just in the performance of mitzvot.

Jupiter: Jesus was a tzedek, righteous and just. (mitzvot not just laws but commandments from God)

Jupiter crowned Saturn (which to the Magi was the star of the Jews) during **all three** conjunctions in 7 BC.

One who was born under the influence of Saturn will be a man whose thoughts are for naught. And some say that everything that others think about him and plan to do to him is for naught.

Saturn: Jesus' message, His thoughts, His words were not accepted by the Jews.

What the Jews planned to do to Him—to rid themselves of Him forever—was unsuccessful. They did not accomplish what they set out to do. They did not kill Him, for He rose on the third day. They did not silence His message.

Chapter 24

Historical Records and Dates

Calendars

Since astronomical event of the Triple conjunction of 7 BC can be identified with all the characteristics described in Matthew along with other historical manuscripts and archaeological evidence, why has there been a problem with the date of the Nativity?

The practice of counting years according to Christ's birth did not start until the 5th or 6th century. Christian calendars, including the Gregorian calendar we now use, have Christ's birth as the "start-date." But when they created the calendar, they did not know when that was.

What exactly IS a year? It is the amount of time it takes the Earth to orbit the Sun and return to its same position comparative to the background stars. It is called the "sidereal year". It does not take exactly 365 years but 365.256 days.

There were several dating systems at the time of Julius Caesar in 44 BC. One was his new Julian calendar, and the other, the Roman Republic Calendar which began with the year Rome was established: 753 BC. To complicate matters, Julius Caesar also determined that a year would be calculated beginning with the start of complete authority by the reigning emperor at that time. (There was also the Hebrew Calendar) In 46 BC. Julius Caesar instituted a new calendar, which became known as the Julian Calendar, it initiated a new dating system in an attempt to realign the calendar with the spring equinox and the solar year. The Julian Calendar had 365 days with an extra leap-year day added every four years and was closer to the solar calendar's length of 365.2422 days. But in order to introduce the new calendar, Caesar added three months into the current year to bring the Roman

Empire's calendar in line with the current solar calendar. As a result, 46 BC became the longest year ever with a total of 445 calendar days.

If Jesus was born in 7 BC, why is our calendar so far off?
The blame lies with a Roman Monk Dionysius Exiguus.

In AD 525 a Roman scholar monk, Dionysius Exiguus, fixed the AD origin of our present calendar (Anno Domini = in the year of our Lord) Hundreds of years after the birth of Jesus, he tried to calculate the exact date of Christ's birth. He chose December 25th because it was already an important date of feasting and marked the time of year of Northern Winter when the Sun was lowest in the sky the Winter Solstice. Dionysius based his extremely complicated calculations on the dates of Easter for the 95 years from his own year of AD 525, during this time he thought he could recalibrate the entire dating system of the whole Christian world. Emperor Diocletian was a persecutor of Christians, and the previous dating system of AD had meant Anno Diocletian the year of Diocletian, he wanted it to start with the year of the Lord, Anno Domini, he then attempted to calculate the year of the birth of Jesus. He supposed Jesus was born 753 years after the foundation of Rome. He was wrong and some of his mistakes are known which meant he ended up 4 to 5 years off. He wrongly dated the birth of Christ according to the Roman system as Dec. 25, 753.

The Anno Domini dating system is used to number the years today in the Gregorian calendar. Scholars discovered that Dionysius was likely off in his reckoning. However, by the time that was discovered, it was too late to change the dating to account for the error. Therefore, we retain our dating system and recognize that the birth of Christ likely occurred between 7 BC and 3 BC rather than in AD 1.

The assertion that the Christian calendar is based on a false premise is not new and many historians believe that Christ was born sometime between 7 BC and 3 BC with the majority of scholars believing 7 BC to 6 BC due the death of Herod being the beginning of 4 BC.

In 590 AD, Pope Gregory I, later known as Gregory the Great, attempted to reform of the Julian Calendar. This Julian system, initiated by Julius Caesar in 46 BC, had accumulated outstanding inaccuracies over the centuries, consequently leading to a misalignment with the solar year and religious Holy Days. The Pope wanted to align the ecclesiastical year with astronomical reality. The inaccuracies had caused the date of Easter, the most significant Christian Holy Day to slip away from the spring equinox, this was a huge issue for the Church. Pope Gregory established the Gregorian calendar, it is used today and gets its name from Pope Gregory XIII, who oversaw its creation and installment in 1582.

"The calculation of the beginning of our calendar based on the birth of Jesus, was made by Dionysius Exiguus, who made a mistake in his calculations by several years," Pope Benedict writes in his book. Benedict says he was several years off in his calculation of Jesus' birth date. "The actual date of Jesus's birth was several years before." Jesus was born several years earlier than believed, according to the book by Pope Benedict XVI. Pope Benedict agreed with the story of the Magi that they were tracking a star, and he identifies it as a conjunction between Saturn and Jupiter. (see Appendix: *Calendars and Astronomical Dates*)

The Fifteenth Year of the Reign of Tiberius

Luke 3:1 *Now in the fifteenth year of the reign of Tiberius Caesar, Pontius Pilate being governor of Judea, and Herod (Antipas)being tetrarch of Galilee, and his brother Philip tetrarch of Ituraea and of the region of Trachonitis, and Lysanias the tetrarch of Abilene,* (KJV)

Luke 3:21 *When all the people were baptized, it came to pass that Jesus also was baptized; and while He prayed, the heaven was opened. 3:23 Now Jesus Himself began His ministry at about thirty years of age,"* (NKJV)

Luke writes Jesus' baptism was in the 15th year of the reign of Tiberius.
Luke 3:1 states, "In the fifteenth year of the reign of Tiberius Caesar," when John baptized Jesus. He was about thirty years old when He began His ministry. There have been discrepancies in dating the fifteenth year. Some scholars, influenced by the work of others, insist on a crucifixion date of AD 33, counting backward and attempting to force the timeline to fit their preconceived ideas, rather than adhering closely to Scripture and historical evidence.

"Conjectural padding to make the dates work? Collecting and reworking information to fit their speculative predetermined interpretations of Scripture forcing it into their own conclusion in order to reconcile the gaps and conflicting data." "Associates for Biblical Research What was the 15th year of Tiberius."

Historians on the Fifteenth Year of the Reign of Tiberius
Although some have used the date when Augustus died in AD 14, as the date to calculate from, evidence shows that Tiberius was co reigning with Augustus before his father's death in AD 12.

According to the ancient Historian Suetonius.
"After two years he [Tiberius] returned to the city from Germania (AD 12) and celebrated the triumph [for his military victories in Germany and Pannonia] which he had postponed, accompanied also by his generals, for whom he had obtained the triumphal regalia....Since **the consuls caused a law to be passed soon after this that *he should govern the provinces jointly with Augustus*** and hold the census with him, he set out for Illyricum on the conclusion of the lustral ceremonies [which culminated the census]; but he was at once recalled, and finding Augustus in his last illness but still alive, he spent an entire day with him in private." (*Augustus* 97:1; *Tiberius* 20–21,)

"It was at this time that I became a soldier in the camp of Tiberius Caesar, after having previously filled the duties of the tribunate. For, immediately after the adoption of Tiberius, I was

sent with him to Germany as prefect of the cavalry. Succeeding my father in that position, and for nine continuous years as prefect of cavalry or as commander of a legion I was a spectator of his superhuman achievements, and further assisted in them to the extent of my modest ability. I do not think that mortal man will be permitted to behold again a sight like that which I enjoyed, when, throughout the most populous parts of Italy and the full extent of the provinces of Gaul, the people as they beheld once more their old commander, who by virtue of his services **had long been a Caesar before he was such in name,** *congratulated themselves in even heartier terms than they congratulated him."* Roman History by C. Velleius Paterculus *Res Gestae Divi Augusti*

"After he had broken the force of the enemy by his expeditions on sea and land, had completed his difficult task in Gaul, and had settled by restraint rather than by punishment the dissensions that had broken out among the Viennenses, **at the request of his father that *he should have in all the provinces and armies a power equal to his own, the senate and Roman people so decreed.*** For indeed it was incongruous that the provinces which were being defended by him should not be under his jurisdiction, and that he who was foremost in bearing aid should not be considered an equal in the honour to be won. On his return to the city, he celebrated the triumph over the Pannonians and Dalmatians, long since due him, but postponed by reason of a succession of wars." *Res Gestae Divi Augusti*

Clearly Tiberius did receive authority equal to that of Augustus and began to reign along with Augustus while Augustus was still alive. It is a well-known fact, that he had become **co-regent** with his ailing father **two years earlier in AD 12.** In that year, he was made supreme military commander over Caesar's armies and provinces, documents attest to his reign being fully in force from that time. Thus, his inauguration in AD 14 as emperor was **only a formalization** of a reign that had begun two years earlier.

"The fifteenth year of the reign of Tiberius would depend on Luke 3:1 What Calendar did Luke use? In spite of the benefits of the Julian Calendar in civil affairs, the apostles in all of their time indications within the text of the New Testament never used the Julian calendar indications. They consistently applied the normal Jewish means of time reckoning,"
The New Testament Calendar
by Ernest L. Martin, Ph.D., 1996
Associates for Scriptural Knowledge

Ernest L. Martin, in *The New Testament Calendar*, asserts that **the apostles (and by extension, New Testament writers) did not use the Julian calendar** but instead used **Jewish time reckoning** for dating events.

However, Luke was not an apostle and Luke states clearly: "Many have undertaken to draw up an account of the things that have been fulfilled among us, **just as they were handed down to us by those who from the first were eyewitnesses and servants of the word.** With this in mind, since **I myself have carefully investigated everything from the beginning**, I too decided to write an orderly account for you, most excellent Theophilus…"
— *Luke 1:1–3 (NIV)*

Luke was **not an eyewitness**. He **investigated carefully** and **interviewed** those who were (likely Mary, the apostles, and others).
Luke was not Jewish — he was **a Gentile**, physician, historian, and traveling companion of Paul, almost certainly **Greek** (see Colossians 4:10–14, where Paul distinguishes Luke from his Jewish co-workers).
Luke was **educated** and **wrote in refined Greek**, showing familiarity with **Hellenistic and Roman chronology**.
Luke wrote both the *Gospel of Luke* and *Acts of the Apostles* primarily for a **Gentile audience** ("most excellent Theophilus" — Luke 1:3), which further supports the likelihood that he followed

Roman (Julian) dating conventions when referring to emperors or civil rulers.

The "fifteenth year of Tiberius Caesar" (Luke 3:1) **This is a Roman imperial reference, not a Jewish one.**

When Luke says "the fifteenth year of the reign of Tiberius Caesar," he's clearly using **Roman imperial regnal years**, not a Jewish calendar formula.

Roman historians (Tacitus, Suetonius) and administrators dated reigns **according to Julian years**.

- The **Julian calendar** had been in use since 45 BC and was well established by Luke's time.
- Jewish calendars were **lunisolar**, but Jews didn't number years by foreign emperors; they used months like Nisan, Iyyar, etc., and festivals (Passover, Tabernacles).

Thus, when Luke ties an event to a **Roman emperor's reign**, he is **clearly using the Roman (Julian) system**, not the Jewish one.

Conclusion

The quote from Ernest L. Martin is **not accurate when applied to Luke**.
While Jewish writers (like Matthew or John) sometimes used Jewish festival-based dating, **Luke**, a **Greek Gentile historian writing for Gentiles**, would absolutely have used **the Julian calendar** and **Roman regnal reckoning** in Luke 3:1.

History and Background: The Jewish calendar is based on a history of time reckoning efforts dating back to ancient times. Both Israelite and Babylonian influences played an important role in its development. The Jewish or Hebrew calendar is a lunisolar calendar created and used by the Hebrew people. It is "lunar" in that every month follows the phases of the moon, and "solar" because the calendar's 12 months follow the earth's orbit around the sun. In parallel with the modern Islamic calendar, the timing of the months in the early forms of the Jewish calendar depended on actual sightings of the Crescent Moon. However, this practice was gradually changed, and by AD 1178 the calculation of the beginning of a new calendar month had been fully replaced by the mathematical approximation of the moment the Crescent Moon begins to appear (Molad) rather than actual sightings.

Luke 3:1 *"In the fifteenth year of the Reign of Tiberius"* does not contradict **John 2:20** on the timing of Jesus' ministry. In actual fact they confirm the timing.

Another Account: Tertullian used the Julian calendar, and there is no discrepancy.

In the much earlier testimony of Tertullian (*Against Marcion* I.xv), it is stated that "the Lord has been **revealed** since the **twelfth** year of Tiberius Caesar." This testimony, generally interpreted as referring to Jesus' baptism and the beginning of His public ministry, when He was indeed "revealed" to the people, comes from Tertullian, who lived from AD 155 to 220. The Julian calendar was in use at the time he made the above statement. If counting from his co-reign in AD 11-12, the **twelfth** year of the reign of Tiberius would have been AD 24. However, Tiberius began his formal reign in AD 14. If counting from and **including** his first formal year of reign, beginning in AD 14, his **twelfth** year would have been AD 25-26. Jesus would have been thirty in the spring and thirty-one years old in the fall of AD 25, as Luke stated, "about thirty years of age." This brings us back to a birth year in September 7 BC. (Jesus' birth in September 7 BC made Him 6 years old in September 1 BC, add 6 to 25 AD =31 years old in September of 25 AD). See page 207.

Jesus was **revealed** at His **Baptism** by John when the Holy Spirit descended upon Him. John the Baptist said *"I myself did not know him, but the reason I came baptizing with water was that he might be **revealed** to Israel."* John 1:31

Date of the Temple
John 2:18-20 (NASB) The Jews then said to Him, "What sign do You show us as your authority for doing these things?" Jesus answered and said to them, "Destroy this temple, and in three days I will raise it up." The Jews then said, "It took **forty-six years** to build this temple, and yet You will raise it up in three days?"

"And now Herod, in the eighteenth year of his reign, and after the acts already mentioned, undertook a very great work, that is, to build of himself the temple of God, and make it larger in compass, and to raise it to a most magnificent altitude, as esteeming it to be the most glorious of all his actions, as it really was, to bring it to perfection; and that this would be sufficient for an everlasting memorial of him; Josephus Antiquities

Herod was given the rule by the Romans under Caesar Augustus in 40 BC, before Antigonus was executed. When Hyrcanus II was captured during a Parthian invasion in 40 BC, however, Herod was forced to flee to Rome to beg for assistance. With the help of powerful friends, such as Mark Antony, Herod was proclaimed king of Judea by the Roman senate and given sufficient military support to reclaim his new kingdom from the Parthians.

"Hereupon Antony was moved to compassion at the change that had been made in Herod's affairs, and this both upon his calling to mind how hospitably he had been treated by Antipater, but more especially on account of Herod's own virtue; so he then resolved to get him made king of the Jews, whom he had himself formerly made tetrarch. The contest also that he had with Antigonus was another inducement, and that of no less weight than the great regard he had for Herod; for he looked upon Antigonus as a seditious person, and an enemy of the Romans; and as for Caesar,"...

"So he called the senate together, wherein Messalas, and after him Atratinus, produced Herod before them, and gave a full account of the merits of his father, and his own good-will to the Romans. At the same time they demonstrated that Antigonus was their enemy, not only because he soon quarreled with them, but because he now overlooked the Romans, and took the government by the means of the Parthians. **These reasons greatly moved the senate; at which juncture Antony came in, and told them that it was for their advantage in the Parthian war that Herod should be king; so they all gave their votes for it. And when the senate was separated, Antony and Caesar went out, with Herod between them; while the consul and the rest of the magistrates went before them, in order to offer sacrifices, and to lay the decree in the Capitol. Antony also made a feast for Herod on the first day of his reign."** Flavius Josephus War of the Jews Book I Chapters 1-20 (Chapters 14:4)

This took place three years before to the death of Antigonus in 37 BC.

"However, the king resolved to expose himself to dangers: accordingly he sailed to Rhodes, where Caesar then abode, and came to him without his diadem (crown), and in the habit and appearance of a private person, but in his behavior as a king. So he concealed nothing of the truth, but spike thus before his face:"O Caesar, as I was made king of the Jews by Antony, so do I profess that I have used my royal authority in the best manner, and entirely for his advantage;

I have laid aside my diadem (crown), and am come hither to thee Caesar replied to Herod "Nay, thou shalt not only be in safety, but thou shalt be a king; and that more firmly than thou wast before;
F. Josephus Book 1 Chapter 20.

In the above excerpt from Josephus, we read that Herod had been made "King of the Jews" by Mark Antony and the Roman Senate. The consensus among historians, based on a careful reading of Josephus and its correlation with Roman history, is that the meeting described above occurred **in the winter of 40/39 BC.**

Here is the detailed timeline that leads to this conclusion:
1. **40 BC:** The Parthians invade Roman Syria and Judea. They capture Jerusalem and install **Mattathias Antigonus**, of the Hasmonean dynasty, as both King and High Priest. Herod's brother, Phasael, is captured and dies. Herod himself is forced to flee Jerusalem with his family.
2. **Herod's Escape and Journey:** Herod flees first to the fortress of Masada and then to Petra (in Nabatea). After being refused significant aid there, he decides to go to Egypt and then on to Rome to seek official recognition and military support from the Roman Senate.
3. **The Meeting in Rome (The Key Event):** Herod sails to Rome. He first met with **Marc Antony**, his powerful patron, who then brought him before **Octavian** (the future Caesar Augustus). Josephus records the speech quoted above, where Herod lays his case before them. Moved by his story and seeing him as a reliable client king to fight the Parthian-backed Antigonus, the Roman Senate officially appoints Herod "King of the Jews."
4. While Josephus's account can sometimes present chronological challenges, the evidence strongly points to the meeting between Herod and Octavian (Caesar) taking place in the winter of 40 and early 39 B.C.E., with Herod being formally declared king by the Roman Senate during this visit.

Let's clarify the sequence and the nature of the titles involved. Herod was not lying, and the titles are not contradictory, but they represent two different levels of authority and legitimacy.

The Two-Stage Process of Becoming King

Stage 1: Nomination by a Powerful Patron (Marc Antony)

After Herod fled Jerusalem in 40/39 BC, he made his way to **Alexandria, Egypt**, where Marc Antony was at the time. **Antony, as one of the Roman Republic's two most powerful men (a Triumvir), officially nominated Herod for the kingship.** This was a massive political endorsement. Think of it

as Antony saying, "You are my candidate. I support you and will provide the military muscle to make this happen."
However, Antony alone could not formally bestow the title. The authority to appoint client kings for major territories ultimately resided with the **Roman Senate**.

Stage 2: Formal Ratification by the Senate (with Octavian's support)

From Egypt, Herod and Antony sailed to Rome.
- Antony, using all his influence, presented Herod to his fellow Triumvir, **Octavian (the future Caesar Augustus)**, and to the Roman Senate.
- This is the meeting Josephus records. Herod's speech is a masterclass in political diplomacy:

- **"I was made king of the Jews by Antony..."** He acknowledges and praises his powerful patron, assuring Antony he has been loyal.
- **"...I have laid aside my diadem (crown)..."** This is a symbolic act of supreme political intelligence. He is saying, "I do not claim this crown by my own right or force. I lay it before the ultimate authority of Rome (the Senate and you, Octavian) to confirm or deny."
- Octavian and the Senate were impressed. They saw a capable, loyal, and politically astute ally who could rule Judea effectively as a Roman client and be a bulwark against the Parthians.
- **Caesar (Octavian) then replied, "...thou shalt be a king; and that more firmly than thou wast before..."** This is the formal, constitutional act. The Roman Senate **ratified Antony's nomination** and officially appointed Herod "King of the Jews."

The Analogy
Think of it like a modern political appointment:
- **Antony's Nomination:** Like a powerful CEO (Antony) telling a candidate (Herod), "The job is yours. I'm backing you. Let's go to the Board of Directors to make it official."
. **The Senate's Ratification:** The Board of Directors (the Senate), with the crucial support of the Chairman of the Board (Octavian), officially votes and appoints the candidate to the position.

According to *Josephus, Antiquities* XV.380, Herod the Great **began** to build Jerusalem's temple **during** the **eighteenth** year of his reign.

The Temple and the age of Jesus

"Forty-six years this temple was being built and You will raise it up in three days?" (John 2:18-20)

In the Gospel of John, chapters one and two, the events are presented in a clear chronological sequence, with verses beginning with phrases such as "the next day" and "on the third day."

In chapter one, after John the Baptist gives his testimony to the priests and Levites from Jerusalem who ask him, "Who are you?" (*John* 1:29), the narrative begins with the phrase, "**The next day,**" marking a deliberate chronological progression. In *John* 1:31, John the Baptist declares, "I did not recognize Him, but so that He would be revealed to Israel, I came baptizing in water." The subsequent passage, *John* 1:35–42, opens with... "Again, **the next day,** John (the Baptist) stood with two of his disciples**; and looking at Jesus as He walked, he said, 'Behold, the Lamb of God!'" One of these two disciples is identified as Andrew; the other disciple is traditionally understood to be John, the author of the Gospel, who presents himself implicitly within the narrative rather than by name. While the text does not explicitly state that "John was the other disciple," the evidence is overwhelmingly circumstantial. According to the Gospel account, the first two disciples were Andrew and an unnamed disciple. Based on the author's literary patterns, the intimate detail, and strong historical tradition, that unnamed disciple is understood to be the Apostle **John**, the son of Zebedee.

The Clue in the Intimate Details: The passage includes a specific, seemingly trivial detail that an eyewitness would remember: **"It was about four in the afternoon."** This kind of personal recollection suggests the author was one of the two men present. Then first thing Andrew does is find his brother Simon Peter *"The first thing Andrew did was to find his brother Simon and tell him, "We have found the Messiah"* John 1:41

The very next verse, 1:43 also begins with "**The next day,**

Jesus decides to go to the Galilee where He finds Philip, who then tells Nathaniel.

This journey fits perfectly with the sequence of "days" mentioned in John 1. The call of the disciples and the journey to Galilee happened *before* the "third day" of the wedding at Cana. **(John 1:40-42):** Andrew finds his brother Peter (who was also from Bethsaida). They are still in the Jordan River region. **(John 1:43-44): "The next day Jesus decided to leave for Galilee."** This marks the start of the 4-day, ~120 km journey north. The group (Jesus, Andrew, Peter, and the other disciple) travels north. It is **during this journey** that Jesus finds Philip (who was from Bethsaida), and Philip finds Nathanael (who was from Cana). These encounters likely happened as they entered the region of Galilee.

John 2:1-2 **On the third day** there was a wedding in Cana of Galilee, and the mother of Jesus was there. Now both Jesus and His disciples were invited to the wedding.

John 2:12 -15 *After this He went down to Capernaum, He, His mother, His brothers, and His disciples; and they did not stay there many days. Now the Passover of the Jews was at hand, and Jesus went up to Jerusalem. And He found in the temple those who sold oxen and sheep and doves, and the money changers doing business.* ***When He had made a whip of cords,*** *He drove them all out of the temple, with the sheep and the oxen, and poured out the changers' money and overturned the tables.*

Another Clue in the Intimate details: John is the only one, when describing the cleansing of the temple event, who mentions the specific point, "when he had made a whip of cords."

John records Jesus purging the Temple after the wedding at Cana and a brief stay in Capernaum. Although the Synoptic Gospels (Matthew, Mark, and Luke) place the cleansing of the Temple at the end of His ministry, the timing shown through

Scripture indicates that Mark, Luke, and Matthew were not eyewitnesses to the event. Some scholars believe that Luke and Matthew derived their accounts from Mark. This is possible; however, Mark was not an eyewitness himself or even a disciple. Early church fathers (Papias, as recorded by Eusebius) describe Mark as Peter's interpreter, who wrote down what he remembered from Peter's teachings. This means **Mark's Gospel is based on testimony**, primarily Peter's, rather than Mark having directly witnessed all of Jesus' ministry. Mark **was not present** for the public ministry, miracles, or teachings of Jesus. Unlike Matthew, John, or the other disciples, Mark **was not present** for Jesus' ministry events. His Gospel is considered a **careful secondhand account**, shaped by Peter's eyewitness testimony.

Luke was a Gentile and historian and not one of Jesus disciples. Matthew was not present during Jesus' first miracle at the wedding in Cana or the cleansing of the Temple. The Sermon on the Mount precedes the calling of Matthew, because he was called after several early miracles and teachings, which include the Sermon on the Mount (Matthew 5–7). After Jesus healed the leper (Matthew 8:1), came to the region of the Gadarenes and healed the two demon-possessed men (Matthew 8:28–34), and healed the paralytic (Matthew 9:1–8), Matthew the tax collector was called by Jesus from his tax booth (Matthew 9:9).
Based on Scripture, we can see that neither Mark, Luke, nor even Matthew were present during the Temple cleansing recorded by John. This is evident due to the lack of specific details in their Gospels, details found only in John's Gospel, as well as the differences in chronology surrounding the Temple cleansing. We know that John was an eyewitness; he was one of the very first disciples called after the baptism of Jesus.
Mark and Luke were not disciples and not a witness to any events during the life of Jesus. And Matthew was called later from the tax booth, (Matthew 9:9).

While Matthew, Mark, and Luke place the Temple cleansing later, John, an eyewitness, records the event chronologically and clearly at the beginning of Jesus' ministry, when the Judeans responded to Him by saying, **"It has taken forty-six years** to build this temple" (John 2:20). To resolve this issue, some scholars agree there may have been two occurrences. Either way, the Judeans' statement that "it took 46 years to build the Temple" appears **only** in the Gospel of John. Therefore, we begin by addressing this statement and its timeline from John's account.

As John wrote, Jesus attends the wedding in Cana in the Galilee, where He performs His first miracle after His Baptism (John 2:12-20). Then, in John 2:12-20, John places Jesus cleansing the temple during the Passover week of His public ministry, after the wedding in Cana. The temple had been in the process of construction for forty-six years. Calculating from Herod's first year of reign **39 BC**, which was explained in the previous paragraphs and confirmed by Caesar Augustus in the winter of 40/39 BC, the eighteenth year of Herod's reign was late 21 BC. Subtracting 21 years from 46 years of building, places the date for that **Passover AD 26,** which also brings the year of Jesus' Baptism to AD 25/26. In the spring of AD 25, Jesus was still 30 years old and turned 31 years old in September of AD 25. Since He had just turned Thirty-one, He was still Thirty-one at that time in spring of AD 26. This aligns with Luke 3:23 "Now Jesus Himself began His ministry at **about** thirty years of age."

When adding the factor of the Gospel of John recording Jesus observing even four Passovers in Jerusalem this places the time period for his Baptism in early 26 AD and the crucifixion date April 7, AD 30.

*Now Jesus Himself began His ministry at **about** thirty years of age, being **(as was supposed) the son of Joseph,** the son of Heli,* (Luke 3:23)
"Supposed"?
NIV: "He was the son, *so it was thought,* of Joseph,"
NASB: "being, as was *commonly held,* the son of Joseph,"
CSB: "and *was thought to be* the son of Joseph,"
ISB: "He was (as legally calculated) the son of Joseph,"

Note: An important consideration, many believed Jesus to be the son of Joseph, **"as was supposed,"** and not the Son of God. They would have considered His birth date to have been later, believing He had been conceived by Joseph after his marriage to Mary later, by this calculation they considered Him to be "about thirty." However, earlier, while Joseph and Mary were betrothed, Jesus was conceived by the Holy Spirit, Matthew 1:18-21; Luke 1:26-45), six months prior—when Mary visited Elizabeth shortly after conception, while she was not yet married to Joseph.

Jesus would have turned 31 years old the September before that Passover (AD 26) and was still 31 at that time, in stead of 30 which was calculated from the actual **marriage** of Joseph and Mary, not His conception by the Holy Spirit and earlier birth in September AD 7.

Let's examine the calculations using Scripture, astronomical data, and historical records.

To determine the date of Passover in **AD 26**, we need to consider the Jewish calendar, which is a lunar calendar, and how it aligns with the Julian calendar used at the time. **Passover begins on the 15th day of Nisan** in the Jewish calendar.

Nisan 1 in AD 26 corresponds approximately to March 24, AD 26 (Julian calendar). However it falls on March 22, AD 26 on the Gregorian Calendar. (Verified on astronomy program Starry Night and is based upon the Gregorian Calendar).

Therefore, Passover **(Nisan 15)** would begin at sundown on April 7, AD 26, and continue through April 8, AD 26. It was April 5, AD 26 **on the Gregorian Calendar.** (Verified astronomically). **Passover in AD 26:** Sundown **April 5** to sundown **April 6**, Gregorian calendar.

To understand why people believed Jesus was about 30 years old, rather than His actual 31 (having turned 31 six months earlier), we must examine the events that took place.

Elizabeth was six months pregnant at the time, as the Angel

Gabriel said to Mary, in (Luke 1:25-26). Mary went to visit Elizabeth her cousin and stayed with her until after John the Baptist was born and his *Brit Milah* (circumcision), when the naming of the child took place. *When Zechariah was asked the name of the child, he asked for a writing tablet, and wrote, saying, "His name is John."* So they all marveled. Luke 1:63

Mary and Joseph could not have legally married until Mary returned to her home in Nazareth after her visit with Elizabeth, following the birth of John the Baptist the around March 31,7 BC (a date calculated to align with the priestly divisions and the conception of Jesus announced by the Angel Gabriel). Mary **remained with Elizabeth through John's circumcision** on the eighth day. That year, the Feast of Passover fell on April 5–6, 7BC, according to the Gregorian calendar. Mary would not have left until after Elizabeth completed her **forty days of purification.** Forty days after March 31 is May 10. Allowing time for Mary's travel back to Nazareth, Joseph then realized that she was pregnant. After being instructed in a dream not to put her away but to take her as his wife, and depending on the length of time between Mary's return after May 10 and Joseph's encounter with the angel, their marriage would likely have taken place in mid- to late June, or possibly early July, 7 BC. Calculating from their **legal marriage**, this helps explain why people at that time believed Jesus was "**about** thirty years old" in the **fall of AD 25 and the spring of AD 26,**" having *supposed* Him to be the son of Joseph.

Chapter 25

The Birth of John the Baptist

Luke 1:5 " In the days of Herod, king of Judea, there was a priest named **Zechariah, of the division of Abijah;** and he had a wife from the daughters of Aaron, and her name was **Elizabeth**".

Luke 1:13 "But the angel said to him,"Do not be afraid, Zacharias, for your prayer is heard; and your wife Elizabeth will bear you a son, and you shall call his name **John**".
Luke 1:23-24 **"When the days of his priestly service were concluded,** he went back home. Now after those days his wife Elizabeth conceived; and she hid herself five months,".

Luke 1:24-28, 31 chronicles these events, "Now after those days his wife **Elizabeth conceived;** and she hid herself five months, saying, 'Thus the Lord has dealt with me, in the days when He looked on me, to take away my reproach among people.' Now **in the sixth month** the angel Gabriel was sent by God to a city of Galilee named Nazareth, to a virgin betrothed to a man whose name was Joseph, of the house of David. The virgin's name was Mary. And having come in, the angel said to her, 'Rejoice, highly favored one, the Lord is with you; blessed are you among women!'...And behold, **you will conceive** in your womb and bring forth a Son, and shall call His name Jesus."

Luke 1:36 "And behold, even your relative **Elizabeth** herself has conceived a son in her old age, and she who was called infertile is **now in her sixth month**".

The Exodus Hebrew Calendar inter calculation pattern perfectly matches up with the starting point for counting back the priestly divisions to get to that of Abijah, when Zacharias was serving in the Temple (Luke 1:5–25).

Based on the biblical record and the order of the priestly divisions outlined in **1 Chronicles 24:1–19**, we can estimate that **Jesus was born around the month of Tishri**, which falls in **mid to late September**. This approximation begins with the conception of **John the Baptist**, which took place during the month of **Sivan** (late June - July in 8 BC). After Zechariah's priestly service, Elizabeth likely conceived shortly after, and **John's birth** would have occurred near the **end of March in 7 BC**. Counting **six months forward from Sivan** (end of June/July) brings us to **Kislev** (December), the time of **Gabriel's announcement** to Mary concerning the conception of Jesus. From there, counting forward nine months—the normal period of human gestation—brings us to **Tishri (September)**, the time of Jesus' birth. This timing aligns with the **third conjunction on September 27, 7 BC**, and with **Rosh Hashanah**, supporting it as the likely date of **Jesus' birth**.

Zechariah and the Priestly Divisions

- **1 Chronicles 24:1–19** explains how King David **organized the priests into 24 divisions (courses)**, assigning them duties in the Temple.

- Each division (or course) was headed by a chief priest, and each course served in the Temple **one week at a time**, rotating through the year.

- **Zechariah, the father of John the Baptist, belonged to the 8th division, Abijah** (1 Chronicles 24:10).

The **chronological and weekly rotation of priestly service** is derived directly from this passage.

Weeks are approximate, assuming **weekly rotations from Sabbath to Sabbath** starting at **Passover**. After week 24, the rotation repeats for the rest of the year, with some courses serving **partial weeks for special feasts** like Passover, Pentecost, or Tabernacles.

Let's carefully unpack it step by step.

1. **Priestly Courses and Zechariah**: Zechariah was in the **8th priestly division** (Abijah). Each course served **one week at the Temple**, beginning on **Sabbath, from Saturday to Friday**.
2. **Starting Point**: If we begin counting **from Passover in 8 BC**, that's week #1. Each successive week corresponds to the next course.
3. **Timing of Pentecost**: Pentecost (Shavuot) occurs **50 days after Passover**, which is **7 weeks after Passover**, so it lands in **week 8** in theory. But here's the nuance:
    - The Temple course weeks are counted **from Sabbath to Sabbath**, not strictly from the first day of Passover.
    - If Passover begins mid-week, the first "week" of a course might not align exactly with the calendar weeks.
4. **Implication**: If Pentecost falls **in the middle of the 8th course's week**, the priestly service for that course would overlap with Pentecost. But it wouldn't automatically push it into the 9th week unless the calendar counting causes a partial week overflow.

So essentially: counting from Passover in 8 BC, the 8th course would still mostly correspond to the week when Pentecost occurs, but careful attention to exact Sabbaths and calendar alignment is needed to confirm whether any part of Pentecost falls into the 9th week.

Determining the day of Passover in 8 BC

Lunar calendar mechanics: Nisan is lunar, so it begins with the new moon (molad). The first new moon that occurs after the spring equinox sets Nisan 1. This ensures that Passover, Nisan 14 always falls in the spring.

1. **Nisan and the Spring Equinox**
   - Spring equinox of 8 BC was March 20/21 (Verified by Astronomically- Gregorian Calendar)
   - The first new moon just after the spring equinox was April 13th in 8 BC. This sets the beginning of **Nisan 1 = April 13th, 8 BC.**
   - To find Passover of that year, counting from Nisan 1 to **Nisan 14** (when Passover is celebrated) gives us the date of Passover in 8 BC at sunset **April 27, 8 BC** for Passover to begin.

   Zechariah's Week
   - Zechariah's priestly division is Abijah and is the 8th division.
   - Each division is one week. 7 x 8 = 56 days
   - Counting 56 days from Passover on April 27, is sunset June 22, 8 BC
   - We know that the first Priestly week must begin the **first Sabbath after Passover,** we do not know exactly how many days after Passover it fell. Was the next Sabbath one day, two or three? It cannot be longer than a six days.
   - We also know it cannot be any sooner than 56 days (or 8 consecutive weeks to be the eighth division).
   - In any given year, including 8 BC, the Sabbath could fall a few days to six days after Passover. Therefore, the first priestly division following Passover on **April 27** could have begun on **April 27th or up to six days later.** This would place the **eighth priestly division of Abijah**, to which Zechariah belonged, ending approximately a week later, about **June 28, 8 BC** on the Gregorian calendar, (or **June 26** on the Julian calendar).

- Therefore confirmed by God's calendar in the sky astronomy, and first new moon after the spring equinox of 8 BC. **Zechariah's eighth week would have ended** the earliest **June 22 to June 28.**
- The calendar places the end of Zechariah's priestly division in the **last week of June,** with Elizabeth conceiving shortly thereafter anytime from June 22 into the month of **July of 8 BC.**

John the Baptist Birth Date.

John could only be conceived after the end of Zechariah's priestly term which as was determined that year to be June 22 to June 28, 8 BC.

- From **conception - True biological gestation** 38 weeks or 266 days.
- Based on when Zechariah's priestly term ended, the earliest his wife Elizabeth could have conceived is from June 22 to June 28.
- Counting 266 days from June 22 and June 28, we arrive at the earliest birth date for **John the Baptist** to be **March 20, 7 BC to March 24, 7 BC.**

The Angel announced to Mary that "Elizabeth is six months pregnant". Therefore, that angel's visit to Mary, must be at least six months later, December 22, 8 BC.

- Using medically confirmed human biology normal fetal gestational days.
- The angel told Mary *behold, you will **conceive** in your womb and bring forth a Son,* (Luke 1:31). Using normal (full-term) fetal **gestational** days brings us to September 27, 7 BC.

- Fetal gestational age ranges from 39 to 40 weeks or **273 to 280** days.

Birth Date: September 27, 7 BC

Gestation Days	Conception Date
280 days	December 21, 8 BC
279 days	December 22, 8 BC
278 days	December 23, 8 BC
277 days	December 24, 8 BC
276 days	December 25, 8 BC
275 days	December 26, 8 BC
274 days	December 27, 8 BC
273 days	December 28, 8 BC

Fetal gestational days and **conceptional age are different.**

Since the angel told Mary "you will conceive", we will now use **conceptional age** (fertilization age).

- From **conception - True biological gestation** 38 weeks or 266 days.

- 266 days brings us to a **conception date of January 4, 7 BC** and **Jesus birth date to September 27, 7 BC.**

This corresponds to about 280 days (40 weeks) when measured from the LMP, because ovulation and conception occur about 14 days after the last period. This again brings us to a conception date during the last week of December to January 4, 7 BC, and a birth date of **September 27, 7 BC.**

Also 14 days after the angel's announcement, Jesus' conception date again calculates to January 4, 7 BC and

Birth Date of Jesus September 27, 7 BC

* Dates are from the Gregorian Calendar as a reference point, because it is the calendar we use today. Julian calendar adjust back two days.

Christmas and Hanukkah

The first night of Hanukkah, which begins on the 25th day of the Jewish month Kislev, changes year to year because the Jewish calendar relies on a lunar calendar, based on how long it takes the moon to go around the Sun. The lunar calendar is about 354 days long, as opposed to the 365-day solar calendar, which is the time it takes for the Earth to revolve around the Sun.

    That shorter lunar calendar means that from one year to the next, Hanukkah which lasts eight nights, will start 11 days earlier. Hanukkah can begin as early as Thanksgiving, However, the Festival of Lights always happens in the winter because Jewish tradition requires holidays to take place during particular seasons.

Though celebrated by different faiths, Hanukkah and Christmas, are both luminous reminders of our shared hope and prayer for a world made whole by peace and goodwill for all.

We are called to bring light into the world. Christmas is a season dedicated to illuminating the darkness with love and kindness.

"Glory to God in the highest,

And on earth peace, goodwill toward men!"

Luke 2:14

Chapter 26

Tracing the Birth-to-Crucifixion Timeline through Scripture and Historical Evidence

In this chapter, we will revisit several key points discussed earlier, such as "The Fifteenth Year of the Reign of Tiberius" and expand upon them with additional evidence. By bringing these pieces together, we can form a complete and chronological timeline from the birth of Christ to His crucifixion.

It is important not to force-fit information to conform to a popular theory. Too often, this subject is approached with a predetermined date already established, largely due to tradition and the assumption that Jesus was born in 1 BC, AD 1, or even 0. The common line of reasoning is that Jesus was baptized at age thirty (Luke 3:21), and since there were three Passovers recorded in the Gospels, He must have been thirty-three years old at the time of His crucifixion, placing the event in AD 33.

However, those who cling to this date often overlook a key chronological fact: if Jesus was crucified in AD 33, He would have been at least 37½ years old—not 33. Why?

First, we must examine both scriptural and historical evidence to establish the key dates that serve as chronological bookends for the period in which the crucifixion could have occurred.

First Bookend: Herod died in 4 BC This is the consensus among scholars and is supported by Scripture, history, archaeology, and astronomy. He did not die in 1 BC, as a some claim. The assertion that "Luke got it wrong in his Gospel" is not valid evidence but rather an expression of frustration at the absence of supporting data.

Josephus informs us that Herod died between a Lunar eclipse (blood moon), and shortly before Passover. Was there a lunar eclipse during that time? Yes — in both 1 BC and 4 BC — but it is important to note that lunar eclipses occur almost every year. In fact, the average number of blood-red moons ranges from two to four, and occasionally even five per year.

Our solar system works like a clock with gears. How often does it strike 3 o'clock each day? Twice — once at 3 a.m. and once at 3 p.m. And when it is noon, we can predict that in three hours it will be 3 o'clock. That is why postings on the internet predicting the next 'blood moon' (lunar eclipse) are common; it is easy to forecast, since the Moon follows its normal orbital path, as it has done for thousands of years.

Because lunar eclipses are relatively common, this information alone is insufficient to determine Herod's death date. Additional historical and astronomical data are required to confirm whether it occurred in 1 BC or 4 BC.

Evidence supports Herod's death date in 4 BC

Josephus (*Antiquities* 17.9.3; *The Jewish War* 2.1.3) records that Herod died between a lunar eclipse and shortly before a Passover. A lunar eclipse occurred on March 13, 4 BC.

Some have accused Josephus of contradicting himself, yet a closer examination shows he does not. Josephus records that Herod reigned for thirty-seven years from his appointment in 40 BC, and thirty-four years from his conquest of Jerusalem in 37 BC (*Antiquities* 17.8.1; *Wars* 1.33.8). Using inclusive counting—a common practice in ancient chronology—this places Herod's death in 4 BC The span of Herod's reign simply does not support a 1 BC date for his death.

References: Josephus, *Antiquities* 17:221–294; Josephus, *Wars* 2:16–74.

The actions of Herod before his death: *When we had done these things, he died, the fifth day after he had caused Antipater to be slain; having reigned , since he had procured Antigonus to be slain, thirty-four years; but since he had been declared king by* *the Romans, thirty-seven."* Flavious Josephus , *Antiquities of the Jews* 17 8:1

Herod was born in 74 BC and was in his seventieth year of age when he died during the *Golden Eagle Event* and the lunar eclipse, confirming that his death occurred in 4 BC.

The **Golden Eagle incident** is one of Josephus' key narratives about unrest just before Herod's death in **4 BC.**

Josephus records this episode in two places—*Antiquities of the Jews*, Book 17, Chapter 6, Sections 2–4 (17.151–163). A group of Jewish teachers, along with their students, cut down the golden eagle that Herod had placed over the Temple gate because it violated the Law concerning graven images. They were arrested, tried, and executed by burning.

The Jewish War, Book 1, Chapter 33, Sections 2–4 (1.648–655), gives a parallel account of this event—depicting the surge of public outrage against Herod and the religious motivation of those who tore down the golden eagle. Their act was not one of rebellion alone, but of devotion, driven by a deep reverence for God's law and a refusal to tolerate graven images within the Temple.

Why It Matters

This event occurred shortly before Herod's final illness and death (*Ant.* 17.6.5; *War* 1.33.8). Josephus places it in direct connection with Herod's rapid physical decline—his failed treatments at Jericho, and his notorious command to assemble the Jewish leaders in the Hippodrome, intending that they be executed upon his death so that mourning would be guaranteed throughout the land. Though that gruesome order was never carried out, the account makes clear that Herod was near death at this time.

This timing firmly anchors Herod's **death in 4 BC**, as Josephus places the Golden Eagle incident, Herod's final illness and death, and the subsequent Passover revolt in immediate chronological sequence.

Herod was born in 74 BC, Josephus writes that Herod died at the age 70 yrs of age. 74-70 = 4

Sabinus, the Roman procurator and treasurer of Augustus:
After Varus had returned to Antioch, between Passover and Pentecost of the year 4 BC, Sabinus arrived at Caesarea, having been sent by Augustus to make an inventory of the estate **left by Herod upon his death.**

*"But Sabinus, Caesar's procurator in Syria, **after the feast was over**, hurried to Jerusalem."* Flavius Josephus *Jewish Antiquities* (Book 17, Chapter 9, 1): and with more detail *The Jewish War* (Book 2, Chapter 3)

The Passover of 4 BC began at sundown on **Monday, April 11, 4 BC** (with the 1st of Nisan beginning the evening before). **Fifty days later** leads to the Feast of Pentecost. the most probable date for Pentecost (Shavuot) in **4 BC** was: Sundown, Saturday, May 29 – Sunday, May 30 6th of Sivan. (Verified on astronomical program Starry Night)

This is the date Josephus describes a massive rebellion against Sabinus and the Roman garrison erupting in Jerusalem, which led to a brutal Roman crackdown. The uprising caused Caesar Augustus to doubt the capability of Herod's son Archelaus, who had assumed reign over Jerusalem. However, because Herod had specifically designated Archelaus as his successor after his death, Augustus made Archelaus *Ethnarch* (ruler of the people) at that time. His brother Antipas and Philip became Tetrarchs and his sister Salome I Toparch (*Timeline: Flavius Josephus, Antiquities of the Jews, Book XVII, Chapter 7, Section 1)*

Herod's death date of 4 BC is not based solely on Josephus.
Mark Antony, Octavian (later named Caesar Augustus), and the **Roman Senate appointed Herod the Great as King of Judea** following the Treaty of Brundisium, sometime in early to mid-October of 40 BC Within seven days thereafter, Herod sailed back to Ptolemais (Acco) on the coast of Judea. This would have been a short 10–15 day leeward (downwind) voyage, arriving in late

October or early November of 40 BC, in time for Herod to gather an army and begin a campaign against Antigonus and the Parthians (cf. Josephus, *Antiquities* 14.14.4–14.15.1; *Wars* 1.14.4–1.15.4), ultimately capturing Antigonus and Jerusalem in **37 BC, Herod's 3rd year since he appointed King** (by Tishri reckoning).

For further analysis of Herod the Great's 40 BC voyage to Rome and return, see:
The Greek historian, Appian, the Perusine war Appian's account in *The Civil Wars* book 5 1.10-6.52,
The Pact or Treaty of Brundisium Sept/Oct 40 BC
The *Concordia Casinum* Inscription
Virgil's 4th *Eclogue*

Herod's Sons Began their Independent Reigns in 4 BC.
According to Origen and Eusebius, the Holy Family returned from Egypt in the first year of the reign of Herod's son Archelaus, who began ruling in April of 4 BC. Joseph was told in a dream that the one who sought to kill the child (Herod) was now dead. If Herod had died in 1 BC, as some claim, the family would have returned while Herod was still alive. Instead, they returned in the first year of Archelaus' reign, which began in 4 BC, after Herod's death.

Matthew 2:19-22 *Now when Herod was dead, behold, an angel of the Lord appeared in a dream to Joseph in Egypt saying, "Arise, take the young Child and His mother, and go to the land of Israel, for those who sought the young Child's life are dead." Then he arose, took the young Child and His mother, and came into the land of Israel. But when he heard that Archelaus was reigning over Judea instead of his father Herod, he was afraid to go there. And being warned by God in a dream, he turned aside into the region of Galilee.*

Establishing Jesus' Birth Date to Confirm the Crucifixion date.

Matthew 2:1-23 (NKJV) *Now after Jesus was born in Bethlehem of Judea in the days of Herod the king, behold, wise men from the East came to Jerusalem, saying, "Where is He who has been born King of the Jews? For we have seen His star in the East and have come to worship Him."*

The Magi weren't pagan. They were looking for the one born *King of the Jews* and evidently by their methods they were perfectly in line with God, because they were successful in finding Him and came *to worship Him.*

The Gospel records that the Magi first came to Jerusalem and inquired about the *King of the Jews.* This inquiry brought them before Herod the Great, who questioned them carefully regarding the timing of the star. Therefore, **Herod was still alive when Jesus was born.**

This fact is crucial: since Herod's death is historically fixed in **4 BC**, the birth of Jesus must have occurred **before that year**, not later. This eliminates the idea that Jesus was born in 1 BC, because the Magi could not have visited Herod after his death.

Herod died in 4 BC.
1. As written by Josephus Herod died between a blood red moon and Passover.
    (Also verified by Starry Night as March 13, 4 BC)
2. Herod's sons Archelaus, Antipas and Phillip began their reigns in 4 BC.
3. The records of the battle of Brundisium, more evidence proving Herod died in 4 BC.

The only way to conclude that Herod died in 1 BC, or any other year besides **4 BC,** is to ignore the evidence.

With **Herod's death established in 4 BC**, we can now compare this to the traditionally claimed date of Jesus' crucifixion in AD 33. A simple calculation—33 + 4—shows that Jesus would have been at least 37 years old at the time of His death.

We must also account for Herod's decree to kill all male children **"two years old and under"** (Matthew 2:16). This places the birth of Jesus back at least an additional 1½ to 2 years from Herod's death. That means if the crucifixion date was AD.33, Jesus would have been closer to **38½ years old** when He was crucified.

Again, the simple math is clear: AD 33 + 4 BC = 37 years, and when we factor in Herod's decree, the number rises higher.

Unless you can agree that Jesus was at least approximately 38 years old at the time of His crucifixion, then AD 33 is too late.

Archaeological Evidence

Archaeological evidence and tangible historical records are preserved in the words of Caesar Augustus in his autobiography inscribed in stone, *Res Gestae Divi Augusti*. **The census under Caesar Augustus is recorded as having taken place in 8 BC.**

Since Herod's death is firmly established in early **4 BC**, the birth of Jesus must fall between the census of **8 B.C.** and Herod's death in early **4 BC.**

With 5 BC birth being too late for Herod's decree to kill all the male children two years old and under, Herod did not issue the command shortly before his death. Also, a 5 BC Birth is too distant from the 8 BC census for it to have forced Mary and Joseph to travel to Bethlehem while Mary was still pregnant. Roman records indicate that, at times, people from Judea often took a year to comply with census orders. They did not, or were unable to respond to a census until the year following its initial decree. This means Joseph may not have arrived in Bethlehem for the census until the following year after it was ordered by Caesar Augustus. The Cyrene Census Papayrus: The edict was issued in May/June of AD 104, but the census year was recorded as **AD 105**. This provides direct, documented proof that the process of a Roman census could span more than one calendar year, with the **official "census year" being the later one.**

From Rome to the Province: The decree from Augustus in Rome had to travel to the legate of Syria (Quirinius), who then had to organize the logistics for Judea.
Public Proclamation: The order had to be proclaimed throughout the region, giving people time to hear about it and make arrangements.
Travel Time: People like Joseph, who had to travel a significant distance, needed time to arrange their affairs and journey to their ancestral home. The edict from Egypt explicitly acknowledges that people were "away from their homes," proving travel and delay were expected.

With Herod dying in early 4 BC, and given that he ordered the slaughter of male children "two years old and under" in Bethlehem (Matthew 2:6), we must count back to at least two years from Herod's death. And this Census in 8 BC, this places the birth of Jesus earlier than 5 BC—into the later months of 7 BC. And would have left for Egypt January to March 6 BC, (after the visit from the Magi). The Gospel of Matthew indicates that Herod died a significant amount of time after the birth of Jesus, while Jesus and His family were in Egypt for about 2 ½ yrs. Placing their return in the first year of the reign of Archelaus, Herod's son. (Matthew 2:14–16).

The Triple Conjunction of 7 BC

First conjunction May 29, 7 BC - **Second Conjunction September 27, 7 BC** – Third conjunction December 20, 7 BC. The astronomical planets and constellations involved in this triple conjunction are all verifiable with Starry Night software and are confirmed in the Babylonian Talmud to have been viewed as representations of Israel and the *King of the Jews* by the wise men of the time. The Magi—recognized as the most skilled astronomers in their field and coming from a region where Jews, wise men, and astronomers were known to reside—were specifically searching for the *King of the Jews.*
The Triple Conjunction is the most likely candidate for the Star of Bethlehem—not a myth or speculation, but an astronomical event confirmed by Starry Night software and undeniable tangible evidence recorded in Babylonian tablets by astronomers, which are housed in the British Museum today.
The above information provides historical, archaeological, and astronomical evidence that aligns with Scripture, helping to narrow down the time frame for the accurate date of Jesus' birth, thereby establishing the **first bookend.**

The Second Bookend:
The next date to confirm, in order to establish another chronological bookend, is the fifteenth year of the reign of Tiberius Caesar (Luke 3:1). This date is critical because it marks the beginning of John the Baptist's public ministry, which, in turn, allows us to approximate the start of Jesus' ministry and ultimately the time of His crucifixion.

The Fifteenth Year of the Reign of Tiberius.
Luke 3:1 -2 *In the fifteenth year of the reign of Tiberius Caesar— when Pontius Pilate was governor of Judea, Herod tetrarch of Galilee, his brother Philip tetrarch of Iturea and Traconitis, and Lysanias tetrarch of Abilene— during the high-priesthood of Annas and Caiaphas, the word of God came to John son of Zechariah in the wilderness.*

Luke 3:21-23 *When all the people were baptized, it came to pass that Jesus also was baptized; and while He prayed, the heaven was opened.* **3:23** *Now Jesus Himself began His ministry at about thirty years of age,"and the Holy Spirit descended on him in bodily form like a dove. And a voice came from heaven: "You are my Son, whom I love; with you I am well pleased." Now Jesus himself was about thirty years old when he began his ministry. He was the son, so it was thought, of Joseph, the son of Heli,*

Tiberius was the son of Caesar Augustus. When did Tiberius begin his reign?

Luke writes when Jesus' ministry began Luke 3:1 *"the fifteenth year of the reign of Tiberius Caesar"* when John Baptized Jesus, Luke 3:23 *Now Jesus Himself began His ministry at about thirty years of age, being **(as was supposed) the son of Joseph**, the son of Heli,*

Historical Records on the Fifteenth Year of the Reign of Tiberius.

Historical sources indicate that **Tiberius began his co-reign** with Augustus in **AD 11-12**, and his sole reign began upon Augustus's death in **AD 14**. Therefore, the fifteenth year of Tiberius could be reckoned either from AD 25-26 (counting from the co-reign) or from AD 28–29 (counting from his sole reign). This distinction is significant for establishing the timeline of Jesus' baptism—recorded as occurring when He was "about thirty years of age" (Luke 3:23)—and for aligning the Gospel accounts with known historical events. Thus, determining whether Luke's reckoning of Tiberius's fifteenth year begins from his co-reign or from his sole reign provides the **second bookend** for accurately identifying the years of Jesus' ministry and crucifixion.

Although some have used the date of Augustus's death in AD 14 as the starting point for calculation, evidence indicates that Tiberius was co-reigning with Augustus prior to his father's death, beginning in AD 11- 12. Tiberius was granted *imperium* (authority to rule) and shared administrative power with Augustus during the final years of the emperor's life. One scholar has argued that "there is no reliable ancient historical evidence for such a co-regency," and therefore holds firmly to the AD 14 date. However, **ancient historical evidence does exist** and sources do in fact attest to a shared rule between Augustus and Tiberius.

According to the Roman Historian Suetonius, *"He [Augustus] associated Tiberius with him in the administration of the provinces and the general government"* (*The Twelve Caesars: Tiberius*, 20). Suetonius further notes that Augustus "made Tiberius his colleague in the tribunician power for **five years**" (*ibid.*), which demonstrates that Tiberius exercised imperial authority alongside Augustus prior to his death. (Augustus made Tiberius co-reign in power **"five years"** prior his death in AD 14, indicates that Tiberius had imperial authority **as early as 10 BC** with consuls passing the law in **AD 12**.)

"After two years he [Tiberius] returned to the city from Germania **(AD 12)** and celebrated the triumph [for his military victories in Germany and Pannonia] which he had postponed, accompanied also by his generals, for whom he had obtained the triumphal regalia....Since the consuls caused a law to be passed soon after this that ***he should govern the provinces jointly with Augustus*** and hold the census with him, he set out for Illyricum on the conclusion of the lustral ceremonies [which culminated the census]; but he was at once recalled, and finding Augustus in his last illness but still alive, he spent an entire day with him in private." Josephus, *Antiquities of the Jews* 18.1.1 (Whiston's *Augustus* 97:1); Seutonius, *The Twelve Caesars: Tiberius* 20–21. According to the writings of Josephus, Tiberius began his reign while his father, Caesar Augustus, was still alive—he **co-reigned** with him. **We also have records of this from an eyewitness and soldier** who served under Tiberius.

"It was at this time that I became a soldier in the camp of Tiberius Caesar, after having previously filled the duties of the tribunate. For, immediately after the adoption of Tiberius, I was sent with him to Germany as prefect of the cavalry. Succeeding

my father in that position, and for nine continuous years as prefect of cavalry or as commander of a legion I was a spectator of his superhuman achievements, and further assisted in them to the extent of my modest ability. I do not think that mortal man will be permitted to behold again a sight like that which I enjoyed, when, throughout the most populous parts of Italy and the full extent of the provinces of Gaul, the people as they beheld once more their old commander, who by virtue of his services **had long been a Caesar before he was such in name,** *congratulated themselves in even heartier terms than they congratulated him."* Roman History by C. Veleius Paterculus *Res Gestae Divi Augusti*

"*After he had broken the force of the enemy by his expeditions on sea and land, had completed his difficult task in Gaul, and had settled by restraint rather than by punishment the dissensions that had broken out among the Viennenses,* **at the request of his father that he should have in all the provinces and armies a power equal to his own, the senate and Roman people so decreed.** *For indeed it was incongruous that the provinces which were being defended by him should not* **be under his jurisdiction,** *and that he who was foremost in bearing aid should not be considered an* **equal** *in the honour to be won. On his return to the city, he celebrated the triumph over the Pannonians and Dalmatians, long since due him,* **but postponed by reason of a succession of wars**." *Res Gestae Divi Augusti*

The Jewish historian Josephus also indirectly supports this timeline. In Antiquities of the Jews (18.32), he notes that **Pilate was appointed governor of Judea by Tiberius**, a sequence of events that aligns with an earlier start to Tiberius's effective reign. This corroborates the view that Tiberius's rule was already established before AD 14, strengthening the argument for counting his fifteenth year from the co-reign already established in AD 11-12 rather than from his sole rule after Augustus's death.

Since **Pontius Pilate** was appointed governor of Judea in **AD 25/26 -36**, this aligns precisely with the **fifteenth year of the reign of Tiberius** when calculated from his **co-reign beginning in AD 11-12**. This correlation provides strong evidence that Tiberius was already exercising imperial authority **by** AD 25/26 —rather than AD 28, as would be the case if his reign were reckoned solely from Augustus's death in AD 14.

This understanding is crucial when interpreting **Luke 3:1**, which states that *"In the fifteenth year of the reign of Tiberius Caesar... the word of God came to John the son of Zechariah in the wilderness."* If Luke was indeed counting from the co-reign beginning in AD 11-12, then the commencement of **John the Baptist's ministry** occurred in AD 25-26. This date, in turn, places the **baptism of Jesus**—which followed shortly thereafter —also in AD 25-26, when He was "about thirty years of age" (Luke 3:23).

This timeline harmonizes the **Gospel chronology** with **Roman historical records**, reinforcing the view that Luke's account is historically accurate and **culminating in the Passover of AD 30**. This is the most consistent date for the crucifixion of Jesus based on astronomical, historical, and scriptural evidence. Astronomical calculations confirm that **Nisan 14** (the date of Passover) fell on **Friday, April 7, AD 30**, matching the Gospel accounts of the crucifixion taking place before the Sabbath.

This dating aligns with the governorship of Pilate (AD 25/26–36), the priesthood of Caiaphas, and the broader Roman administrative timeline, all of which converge around AD 30. Thus, the co-reign of Tiberius beginning in AD 11- 12 not only clarifies Luke's chronological reference but also provides the most coherent historical framework for the **life, ministry, and crucifixion of Jesus Christ**.

Ancient coins from Antioch dated AD 12 display the head of Tiberius and documents attest to his reign being fully in force from that time.

Coin minted in co-regency period

Coin minted during co-regency period, showing Augustus on one side and Tiberius on the other. Image courtesy of Classical Numismatic Group www.cncoins.com

There were coins of Tiberius beginning in AD 12, and Roman records clearly indicate he had equal power at that time with Augustus.

The 15th year of the reign of Tiberius is an important landmark in determining the date of Christ's baptism (Luke 3:1). Tiberius (right) succeeded Caesar Augustus (left), the Roman Empire's first Caesar and the one who "caused all the world to be taxed" (Luke 2:1). But Tiberius was not the son of Augustus. The two surviving heirs, Lucius and Gaius, died within a year and a half of each other, and Augustus, suddenly without an heir, adopted Tiberius in AD 4, when Tiberius was 46 years old. Ten years later, Augustus died (AD 14), and Tiberius began his sole reign as Caesar.

The year of Augustus' death is firmly fixed and confirmed by a Lunar eclipse recorded by the Roman historian Tacitus (*Annals*, Book 1, paragraph 28).

Acts of Augustus *Res Gestae;* Copies were sent to the provinces and carved in imperishable **stone** on temple walls. The copy from the Temple of Roma and Augustus in **Ancyra (the Monumentum Ancyranum)** is the most complete and famous. Ancyra, Galatia (Stone Wall) Preserved by being carved into a temple wall in a provincial capital, forgotten, and rediscovered in 1555, thus saving the text for posterity. Other fragments have been found at Apollonia in Pisidia and Antioch in Pisidia.

The Account of Tertullian

In the much earlier testimony of Tertullian (*Against Marcion* I.xv), it is stated that "the Lord has been revealed since the twelfth year of Tiberius Caesar." This testimony—generally interpreted as referring to Jesus' baptism and the beginning of His public ministry, when He was indeed "revealed" to the people—comes from Tertullian. Jesus was revealed at His baptism when the Holy Spirit descended upon Him:

> "And the Holy Spirit descended in bodily form like a dove upon Him, and a voice came from heaven which said, 'You are My beloved Son; in You I am well pleased.'" (Luke 3:22)

He was also revealed by John the Baptist:

> "The next day John saw Jesus coming toward him, and said, 'Behold! The Lamb of God who takes away the sin of the world!'" (John 1:29)

> "I didn't know Him, but I came immersing with water so that He might be revealed to Israel." (John 1:31)

Tertullian claims Jesus was "revealed" **during** the twelfth year of the reign of Tiberius. Tertullian lived from AD 155 to 220; therefore, it is likely he was calculating from the year of Tiberius' formal reign, as recorded in official Roman records, **including** the year AD 14. The **twelfth** year of his reign would thus have been "during" the end AD 25. Therefore, this would also agree with a baptismal date in fall AD 25/26 and **a crucifixion date on AD 30.**

Luke used the Roman Dating System

Church tradition tells us that Luke was a converted Gentile, which scholars suggest is the reason Paul introduces him separately in Colossians 4:11–14.

Luke used the Roman dating system which was reignal "In the fifteenth year of the reign of Tiberius". The Roman dating system was based on the beginning of an emperor's reign, not necessarily tied to a specific linear calendar—whether Julian, Essene, or Hebrew.

Luke 3:1 does not contradict **John 2:20** on the timing of Jesus' ministry. In actual fact they confirm the timing.
Luke 23:54 *"Now it was the Day of Preparation, and Shabbat was approaching. The women who had come with Him from the Galilee followed, and they saw the tomb and how His body was laid. Then they returned and prepared spices and perfumes . But on Shabbat (from Friday sunset to Sunday after sunrise) they rested according to the commandment.*
Gospel says it was a Shabbat preparation day that means it was FRIDAY.
No other guesses or speculation necessary, it clearly says in Scriptures it was Friday
If Jesus was crucified in 30 AD and was born in September 7 BC during the Triple Conjunction,
He would have been 35 ½ years old during Passover in April of AD 30.

Scripture provides no evidence of a blood moon occurring during the crucifixion.

The claim that a partial lunar eclipse in AD 33 marks this event is an example of forcing Scripture to fit a predetermined date—taken out of context and unsupported by any biblical witness.
The theory of a "blood-red moon" on the night of the crucifixion is purely speculative. The Gospels record only that the Sun was darkened for three hours, not that the Moon turned to blood. A solar eclipse was impossible, since the crucifixion occurred

during Passover, which always falls on the 14th of Nisan, when the Moon is full.

The assumption that there was a blood moon during the crucifixion arises from Peter's quotation of the prophet Joel in **Acts 2:16–21.** *"But this is what has been spoken through the prophet Joel:'And it shall come to pass* **in the last days, says God,** *That I will pour out of My Spirit on all flesh; Your sons and your daughters shall prophesy, Your young men shall see visions, Your old men shall dream dreams.
And on My menservants and on My maidservants I will pour out My Spirit in those days; And they shall prophesy. I will show wonders* (Greek: anomalies) *in heaven above And signs in the earth beneath: Blood and fire and vapor of smoke. The sun shall be turned into darkness, And the moon* **into** *blood, Before the coming of the great and awesome/dreadful/notable/terrible day of the LORD."*

Joel's prophecy, as found in **Joel 2:28-31** "**And afterward**, I will pour out my Spirit on all people. Your sons and daughters will prophesy, your old men will dream dreams, your young men will see visions. Even on my servants, both men and women, I will pour out my Spirit in those days. I will show wonders in the heavens and on the earth, blood and fire and billows of smoke. The Sun will be turned to darkness and the **moon to blood** before the great and **dreadful day of the LORD."**

This Scripture should not be taken out of context. This clearly is not talking about the crucifixion but "the great and dreadful day the Lord comes." Joel 2:32 *And everyone who call on the name of the Lord will be saved, for on* **Mount Zion and in Jerusalem** (where?) *there will be deliverance, as the Lord has said, among the survivors who the Lord calls.*

The phrase "the moon into blood" **does not depict the crucifixion** but instead points to the coming "Day of the Lord."

Joel Chapter 3: The context of Joel 3 clarifies that this is the time when God judges the earth. Verse 2 states, *"I will gather all the nations and bring them down to the Valley of Jehoshaphat, and I will enter into judgment with them there on behalf of My people and My heritage,* **Israel.***"*

Taking a passage of Scripture out of context simply because it contains the phrase "moon to blood," and applying it to a predetermined date for the crucifixion without any supporting evidence—even within its own context—is a misrepresentation of the text.

It is quite acceptable exegesis to view the mention of the "blood moon" as a yet-unfulfilled portion of Joel's prophecy that did not apply to the present situation. It was part of the broader context of the passage Peter quoted and was therefore included, even though it was not only relevant to the events of Pentecost. From Peter's perspective, he understood these signs to indicate the approaching end-times. Seeing the manifestation of the Holy Spirit and the speaking in tongues, he fully expected the remainder of the prophecy, including the "moon turning to blood" and the outpouring of God's wrath, to follow shortly thereafter.

This portion of the prophecy remains unfulfilled and points to a future time. If the Sun is darkened, it cannot give off sufficient light to cast the kind of umbral shadow necessary to produce a lunar eclipse or "blood-red moon." Peter, in quoting Joel, was referring both to his present experience of the Holy Spirit being poured out and to the future fulfillment of the astronomical and apocalyptic signs he mentioned. This becomes evident when the passage is read within its full original context in Joel.

In the closing verse of Joel's prophecy, the reference is to the salvation procured through the death of the Messiah and offered to all who believe and call on His name. Joel's "moon to blood" prophecy clearly points to the **end times—the conclusion of history—not the date of the crucifixion.**

Astronomically
Technically, it was, however, a full moon due to orbital positioning, since Passover falls 14–15 days after the new moon. With the Moon's synodic period (orbit) being approximately 29.5 days from new moon to new moon, Passover occurred during the full moon **on Friday, April 7, AD 30.**
Passover falls in the spring month of Nisan in the Northern Hemisphere. The first of Nisan was determined by two witnesses who observed the first sliver of the crescent New Moon—closest to the Spring Equinox, when the Barley is ripe (called Avid), and reported it to the High Priest. Nisan 1 was not a fixed date in the old Jewish calendar, and if the New Moon was not seen at dusk on the expected day, Nisan 1 would be declared on the next.

"The Roman historian Phlegon is mentioned by Origen who describes a darkening of the sun in the **eighteenth year of Tiberius on Nisan 14-Passover which would be 30 AD.** A few people have used this Roman history report stating the year of the Crucifixion is 33 AD. They mistakenly start Tiberius' reign in 14 AD after the death of Augustus instead of the start of his joint reign in 11 AD. For example, it is very clear from Scripture that there is a three hour period of darkness between noon and 3:00 p.m. during the event of the Crucifixion that is mentioned in three of the four Gospels. The historians Thallus (around 52 BC), Phlegon (around 137 AD), and Africanus (around 221 AD) make reference to a three hour period of darkness during the 19th year of Tiberius in 30 AD. They attempt to claim it was some unusual form of solar eclipse that lasted about three hours. They recognized this was a difficult explanation since the moon is opposite the sun during the Nisan 14-Passover on April 5 in 30 AD on the Gregorian calendar. Phlegon writes that there was, "the greatest eclipse of the sun and that it became night in the sixth hour of the day (i.e., noon) so that the stars even appeared in the heavens. There was a great **earthquake in Bithynia**, and many things overturned in Nicaea." This was clearly not explained by the sun being covered up by the moon or clouds". *Jim Liles*

The last supper meal Jesus had with His disciples they sang the Hallel

Mark 14:26 *After singing the Hallel, they went out to the Mount of Olives. Gethsemane where Jesus was arrested.*
When is the Hallel sung?
The Hallel (Psalms 113–118) is sung/recited during the **Feast of Unleavened Bread (Chag HaMatzot)**, which begins with **Passover (Pesach)**.

Here's how it works in Jewish tradition:
- **Passover Seder (first night of Unleavened Bread):** Parts of the Hallel are recited/sung as part of the Seder liturgy (**before** and **after** the festive meal).

Was the meal Feast of Unleavened Bread?

The **Feast of Unleavened Bread (Chag HaMatzot)** is a **seven-day festival** that begins the day after Passover starts:
- **In the Bible (Torah):**
    - Begins on the **15th day of the Hebrew month of Nisan** (right after the Passover meal **on the 14th**).
    - Lasts **seven days** (15th–21st of Nisan).
    - During this time, no leavened bread (chametz) is eaten—only unleavened bread (matzah).
    - The first day and the seventh day are treated as **special Sabbaths** (holy convocations, with no ordinary work).

The last supper Thursday night (which they ate **after sunset** which made it Friday) could have been the Passover meal with Friday being both Passover and also preparing for the special

Sabbath which began Friday night and continued through Saturday to sunrise Sunday morning.

The Teacher asks: *Where is my guest room,* **where I may eat the Passover** *with my disciples?', Then he will show you a large upper room, furnished and prepared; there make ready for us." The disciples went out, came to the city, and found just what Jesus had told them. And they prepared the Passover.* Mark 14: 13-16

It was a Passover meal on a Thursday night, but it was considered Friday because it was after sunset.
Matthew 26:17 **Now on the first day of matzah** (unleavened bread) the disciples came to Jesus, saying, "Where do you want us to prepare for you to eat the Passover?" It was that evening Matt 26:20-30 they ate the meal and later that same night Jesus was arrested.
That same night Jesus was arrested and throughout the night and very early morning
John 18:28 *Then they led Jesus from Caiaphas to the Praetorium. They themselves did not enter the Praetorium, so they would not become unclean but might eat the Passover.*

Thursday night after sunset was considered the next day, which was Friday and Jesus and His disciples had eaten the Passover meal that evening prior to being arrested. Being that it was after sunset and in the evening they ate the Passover meal, was the same day (Friday) that night and early morning when He was convicted and crucified.
Verified by Starry Night April 6th 30 AD was a Thursday. The Last Supper Passover Sedar was eaten after sunset that evening. Jesus was crucified the next morning **Friday April 7th, AD 30** at 9 am, He died at noon.
Everything must align with Scripture—without relying on questionable interpretations, and without assuming that reliable historical records are in error.

Tiberius Caesar, Roman emperor—AD 11/12–37
Pontius Pilate, governor of Judea—AD 25/26–36
Herod the Great died in 4 BC
Herod Antipas, tetrarch of Galilee—4 BC–AD 39 (Archaleus, Antipas and Philip were sons of Herod the Great, after his father Herod died in March 4 BC)
Annas ben Seth, high priest—AD 6–15
Joseph Caiaphas, high priest—AD 18–36

Crucifixion Date Possibilities There are only two, 30 AD and 33 AD. These are the only times Passover begins on a Friday.

- **Friday, April 7, 30 AD***
- Tuesday, March 27, 31 AD
- Monday, April 14, 32 AD
- **Friday, April 3, 33 AD**
- Wednesday, March 24, 34 AD
- Tuesday, April 12, 35 AD
- Saturday, March 31, 36 AD

The day of preparation (Friday) was for a Passover that fell on a preparation day before Shabbat (Saturday). The previous evidence presented supports the crucifixion year of AD 30.
The astronomical program *Starry Night* **confirms**
April 6–7, AD 30, as the 14th of Nisan and the date of Passover.

Final proof—no further sacrifices were needed, for Jesus was the ultimate sacrifice, sufficient for the payment of the sins of the world. As He declared, "It is finished"—paid in full.

The evidence of this completed payment, occurring in the spring during Passover of **AD 30,** is also written in the Talmud.

The Talmud Key Evidence the Crucifixion was no later than 30 AD

There is historical evidence to support that Jesus was not crucified any later than Yom Kippur of AD 30. This was recorded in the Talmud by witnesses from the year **AD 30 forward.**

The first time the cord did not turn white was yom Kippur in fall of **AD 30.**

The Talmud

The Jewish Talmud is the authoritative book of Jewish rabbinical history and tradition.

Yom Kippur is the most sacred of Jewish festivals.

Leviticus 23:27; Leviticus 16:8–9; Exodus 12. After the Exodus, the first month became Nissan 1.

The sacrifice: two goats were presented, and lots were cast to determine the scapegoat and the goat that would be sacrificed for the sins of Israel (Leviticus 16:9–10). The goat sacrificed for the Lord was offered as a sin offering.

Leviticus 16:21 The scape goat was sent away into the desert and the other goat sacrificed.

A crimson chord was taken and one piece of it placed on the scape goat, the other piece remained in the Temple. If the crimson chord that remained in the Temple turned white, it was the sign that God accepted the sacrifice. If it remained crimson, God did not accept the sacrifice for the atonement of sins. Isaiah 1:18 *"Though your sins are like scarlet, They shall be as white as snow; Though they are red like crimson, They shall be as wool.*

In the Talmud, the Temple priests recognized that, 40 years prior to the destruction of the Temple, God no longer accepted the sacrifice for the sins of Israel on Yom Kippur.

According to the Talmud (Mas. Rosh haShana 31b), for 40 years before the destruction of the Second Temple, the cord (thread) of crimson or scarlet never turned white but remained red.

The Temple was destroyed in 70 AD. Calculating back 40 years from the destruction of the Temple in 70 AD. Math 70 – 40 = 30 AD.

The first Yom Kippur following Passover AD 30.

1. The **crimson thread** that remained in the Temple on Yom Kippur did not turn white, it stayed red meaning sacrifices were no longer accepted.

2. The **lot for the goats** never fell to the right hand.

3. The **everlasting light** of the westernmost lamp of the Menorah inside the Holy Place was considered a testimony of the presence of God.
Talmud Shabbat 22b, 6–8 states: "It is a testimony to mankind that the divine presence rests in Israel."
The testimony (22b–8) explains that it was the western branch of the candelabrum into which the same quantity of oil was poured as the others, yet it remained lit. The lamp on the end always stayed burning, unlike the other lamps, signifying that God's presence was there. But from that point forward, it began to go out just like the other lamps.

4. The **doors to the Holy of Holies** began to open—and continued opening.
Two massive doors separated the court from the Holy Place, where only the priests were permitted to enter. According to Josephus, these doors stood approximately 49 feet high and 26 feet wide. They were so heavy that no single man could open them; it required the strength of several men. Only the priests were allowed inside, keeping everyone else out. Talmud Mas Yoma 39b 4-6 (confirm this)
God did not accept anymore sacrifices in the Temple after AD 30.

After Jesus' sacrifice in **AD 30** no further sacrifices were accepted by God. According to the Talmud, God no longer accepted sacrifices for atonement on Yom Kippur

Luke 23:44-45 *The Sun was darkened and the veil of the Temple was torn in two*
After AD 30, the Doors of the Temple kept opening by themselves.
The sacrifice for the sins of Israel was made once a year by the High Priest in the Holy of Holies during Yom Kippur. Only the High Priest had access to the Holy of Holies. When the doors were opened, now everyone had access to not only the Holy Place, but the Holy of Holies.

Timeline from Birth to Crucifixion

Considering Scripture, Herod—who died at the beginning of 4 BC—had the male children in Bethlehem murdered up to two years of age. The Gospel of Matthew indicates that Herod died a significant amount of time after the birth of Jesus, while Jesus and His family were in Egypt (Matthew 2:14–16). Therefore, Jesus had to have been born no later than 6 or 7 BC.

Adding the evidence from Scripture and historical writings that Herod's son Archelaus began his reign in the spring of 4 BC, following his father's death, it becomes clear that Jesus must have been born even earlier. Scripture states that the angel instructed Joseph to return to Israel because Herod was dead. However, when Joseph learned that Archelaus, Herod's son, was now on the throne, he was afraid and continued to Nazareth with Mary and the child Jesus.

Placing the birth of Jesus in 7 BC allows Him to have been a toddler upon returning from Egypt, and before that time, old enough for Herod to have sought His life. Otherwise, if Jesus had been born in 6 or 5 BC, He would have been only one or two years old when Joseph brought the family **back** from Egypt after Herod's death and Archelaus' ascension to take Herod's place.

Timeline of Key Events
Surrounding Jesus' Birth and Early Life

Event	Date
Census under Caesar Augustus Recorded in stone in the Res Gestae Divi Augusti	8 BC
Birth of Jesus Triple conjunction 7 BC	September 27, 7 BC
Herod orders the massacre of all male children up to 2 years. Shortly after birth, before and before Herod's death.	Spring 6 BC to March 4 BC
Joseph takes Mary and Jesus to Egypt.	Spring 6 BC to 5 BC
Death of Herod	Between the lunar eclipse of March 13, 4 BC and Passover April 10/11, 4 BC

Explanation:
- Jesus' birth in **September 7 BC** fits between the **8 BC census** and **Herod's death in 4 BC**, allowing Him to be a toddler at the time of the family's return from Egypt.
- By **1 BC**, Jesus would have been approximately **6 years old**.
- Calculating forward to the **Passover of AD 30**, the year of His crucifixion, Jesus would have been **about 35½ years old**.

Astronomical data: In **AD 30**, Nisan 14 fell on **Friday, April 7**.

Interesting thoughts:
The "Testimonium Flavianum" anchors Jesus' crucifixion under **Pilate's authority**, consistent with the Gospel accounts.
In *Antiquities* 18.63–64, Josephus writes a famous (and disputed) passage about Jesus: *"About this time there was Jesus, a wise man... Pilate, at the suggestion of the principal men among us, had condemned him to the cross..."*
Josephus also discusses **John the Baptist** and Herod Antipas' fear of his influence (*Antiquities* 18.116–119). It confirms the **historical overlap** of the Herodian dynasty and John's ministry.

Josephus does not give us the exact year of the crucifixion. What he **does** provide are the historical goal posts : Pilate: **AD 26–36**

Key notes:
1. **Counting co-reign**: Tiberius' 15th year = AD 25/26 → aligns with Jesus' baptism.

2. **Pilate's term**: 25/26–AD 36., anchored by Gratus' end and Vitellius' removal.

3. **Caiaphas**: High priest under both Gratus and Pilate, removed c. AD 36.

Additional sources on death date of Herod.

Origen (c. A.D. 185–253)
Origen, *Against Celsus*, Book I, Chapter 51
"When Archelaus, after the death of Herod, succeeded to his father's kingdom, Joseph, being warned by God in a dream, withdrew with the Child and His mother into the parts of Galilee..." This shows that Origen understood Archelaus' reign as beginning **immediately after Herod's death**.
Eusebius (c. AD 260–340)
Eusebius, *Ecclesiastical History*, Book I, Chapter 6.6–7
"After the death of Herod, when Archelaus, his son, had succeeded to the government of Judea, Joseph, with Mary and the Child, went to dwell at Nazareth in Galilee." Eusebius explicitly

places the family's return in the **first year of Archelaus' rule**, which historically began in **April of 4 BC.**

Both Origen and Eusebius confirm the Gospel account: Joseph returned only after Herod's death, in the first year of Archelaus, i.e., 4 BC.

Chronological Timeline of Jesus' Lifetime and Related Historical Events

Event	Date (BC/AD)	Age of Jesus
Census of Caesar Evidence of this is Recorded in stone by Roman historians Res Gestae Divi Augusti and Luke 2:1-3; coincides with Quirinius' earlier census in the eastern provinces.	8 BC	not yet born
Triple Conjunction of Jupiter- Saturn Verified by Starry Night Represented the sign of Israel and the *King of the Jews*. Recorded on Babylonian Tablets in the British Museum.	May - Dec 7 BC	Birth Sign
Birth of Jesus Second conjunction of Jupiter and Saturn; aligns with shepherds in the fields (autumn) and Luke's chronology. *Luke* 2:7, *Luke* 2:10-12	September 27, 7 BC	0 years Shepherds found Jesus lying in a manger.

Event	Date (BC/AD)	Age of Jesus
Visit of the Magi to Herod in Jerusalem	December 20, 7 BC	
Matthew 2:7 Herod asked the Magi the exact time the star first appeared. The first sighting of the star alone in the east was in July 8 BC (during the census). Note: The Magi and Herod did not know the age of the child only when the first star appeared in its rising.		3 months - 2 yrs. old
The Magi Arrive	December 20, 7 BC	
in Bethlehem that evening during the Third and final conjunction of Jupiter and Saturn. During the darkest night of the year, winter solstice, creating a Zodiacal light over Bethlehem and the place where the child was. (*Matthew* 2:1-12)"And coming into the **house** they found the young child with His mother"		3 months – 2 yrs. old
Flight to Egypt/ Massacre of the Innocents	6 to late 5 BC	
Herod orders death of males "two years old and under, Matt 2:16).		6 months – 2 yrs. old
Death of Herod the Great	March 13, 4 BC	
Josephus, Antiquities 17.6-9; Between the lunar eclipse and Passover Confirmed by astronomical Starry Night		2 ½ yrs. old
Archelaus Begins his Reign	April 4 BC	
Eusebius and Origen: The Holy family returned "in the first year of Archelaus"		2 ½ - 3 yrs. old

Event	Date (BC/AD)	Age of Jesus
Jesus in Nazareth "He shall be called a Narzarene." Holy family arrive back in Israel and find Archelaus has replaced Herod as ruler, Joseph takes Mary and Jesus to Nazareth.	4 BC- AD 26	From Childhood 30-31 yrs. old
Tiberius' Co-Reign with Augustus begins Seutonius and Roman records show co-regency before the death of Augustus.	AD 11-12	17-18 yrs. old
Fifteenth Year of Tiberius The start of John the Baptist's Ministry Luke 3:1; calculated **from** Tiberius' co-regency in AD 11-12	AD 25/26	30- 31 yrs. old
Jesus' Baptism and Beginning of His Ministry (Luke 2:23); Coincides with John's Ministry start. September AD 25, He turned 31 years old.	AD 25/26	"About 30 years old"
Jesus' Ministry Period Includes multiple Passovers (John 2:13; 6:4; 11:55) but duration of ministry unspecified; not necessarily 3 ½ years.	AD 26-30	31- 35 yrs. old
Triumphal Entry (Palm Sunday) Nisan 10 on the Jewish calendar. Fulfillment of Zechariah 9:9	April 2, AD 30	35 ½ yrs. old
Crucifixion of Jesus Passover Preparation Astronomical data confirms Nissan 14	Friday April 7, AD 30	35 ½ yrs. old

(Note: there are **no** Scriptures stating the Jesus was 33 years old at His crucifixion, this is a common miscalculation assuming he was born in 1 BC or Zero.

As evidence confirms **Herod's death date was in 4 BC.** Therefore, no matter which date one chooses as the crucifixion date **AD 30** or **AD 33**, 4 years must be added to that date. Also an additional one to two years to include the order from Herod to kill male children 2 years and under. As a result, you must include the **additional 1-2 years.**

If you hold to the traditional but unsubstantiated crucifixion date of AD 33, then He was 38½ years old!.
33 + 4 + (1-2 yrs.) = 38-39 years old. 38 ½ years old!

If you believe He was crucified in AD 30, as multiple data supports, He was 35 ½ years old.
30 + 4 + at least 1- 2 years.
30 + 4 + (1- 2 yrs.) = 35-36 years old. 35 ½ years old.

Jesus' Birth	Age in September 1 BC	Crucifixion	Age
September 27, 7 BC	6 yrs old	April 7, AD 30	35 ½

Chapter 27

"Common Misconceptions"

VENUS: Her Star Not "His Star"

When people try to describe the Bethlehem Star, they often end up describing more than one star—Jupiter and Regulus, Jupiter and Venus, Venus and Mars, or Venus and the Moon. So which one, exactly, are they claiming is "His star"? If we are to understand the Bethlehem Star correctly, we must also understand the beliefs of the Magi, *the actual witnesses,* as our guide. It is essential to recognize what *they* would consider the motivating and significant sign, based on their own astronomical knowledge. This mission was not determined on a whim; it was a multi month journey involving considerable cost and risk to their security. They fully understood what they "saw" and what it meant.

Venus
According to the Magi, Venus, known in Hebrew as "shining planet", is female. Jupiter crowning Venus at any time would not have been of interest, nor would it have directed the Magi to the Messiah. The Magi came to see "His star" not *her* star, and to worship the "King of the Jews" not queen of the Jews.

It is well known, and most people are familiar with what Venus represented to the Greeks, Romans and Sumerians. Pythagoras a Greek philosopher, mathematician and astronomer, born 570 BC, was one of the first to realize that Venus as an evening star was the same planet as Venus as a morning star. Venus is an inner planet, meaning its orbit is inside that of the Earth's. It is not visible when it is in transition behind the Sun or when it is in front of the Sun (between the Sun and the Earth). In 10 BC, early Greek astronomers called Venus *Hesperus* ("evening star") when it appeared in the western evening sky, and

Phosphorus ("light-bringer") when it appeared in the eastern morning sky. In classical Latin, Venus was referred to as Lucifer, meaning "light bringer" or "morning star", though the morning star was considered sacred to the goddess Venus (or Aphrodite in Greek mythology). She is the goddess of beauty and often called "she who shines."

Venus is actually the Roman name for the Greek goddess Aphrodite, the goddess of love and fertility. In Roman mythology, it is evident that the Romans personified the stars and planets by naming them after their deities, including Venus, and worshiped them, which is strictly forbidden in Scripture.

To Sumerians, Venus was Inanna/Ishtar, the goddess of love and war, represented by the appearance of Venus as both a morning and evening star. Ishtar, along with being celebrated as the goddess of love, was also associated with sexual attraction and fertility, and was thought to play a crucial role in agricultural cycles and the growth of crops.
To the Magi, Venus wasn't a virgin. Talmudic Hebrew also calls Venus **Kôkebet**, "planetess". Analogous Aramaic formations, **Kawkḫā**, or **Kawkaḇtā**, were generally current. (see sources)

It is common among Christians to use a comparison between Venus, a bright morning star, and Jesus, who in Revelation 22:16 said He is "the Bright and Morning Star". Consequently, some believe the planet Venus represents Jesus at His birth. However, Jesus is not claiming to be represented by Venus, nor is He drawing attention to the planet as a symbol of His birth. When kept in context, at *the end* of the Book of Revelation, the verse refers to the return of Messiah Jesus, the fulfillment of Scripture, and the declaration that the Lord God will be triumphant over all the evil in the world.
Revelation 22:16 "I, Jesus", He is the one speaking: "I am the Root and the Offspring of David, the Bright and Morning Star." Jesus is expressing that He is both the Son of David and, in His

divine nature, Lord God Almighty, the Lord of Hosts, King of Kings, and the Light of the World.
- Venus is not just a morning star; half of the year it is an **evening star.**
- The Book of Revelation, where the title "Bright and Morning Star" was mentioned, had not even been written at the time of the Magi.

The focus, and most important source, when considering the Bethlehem Star is the testimony of witnesses: **the Magi** in Matthew 2:2. The points mentioned above are just a few of the many reasons the Magi would not consider Venus, in any conjunction, to be "His star" or connected to the birth of the "King of the Jews".

What do we know about Venus that the Magi did not know?

We know Venus is a planet, like Earth, not an actual star like the Sun. To the Magi, the difference was, most of the stars were fixed in their positions (*stationary*), while five were not: Mercury, Venus, Mars, Jupiter and Saturn. These were visible to the naked eye, and known as "wandering stars"(*planete* in Greek). The Magi understood these moving lights as the hands on God's great celestial clock.

Stars are primarily composed of hydrogen and helium and give off their own light. Venus is not technically a star, it does not give off its own light, but reflects light from the nearest star, our Sun, for the same reason our Moon does: Sunlight reflects off the thick clouds of Venus's atmosphere.

"Through photosynthesis, we observe the pivotal role the Sun plays in sustaining life on Earth. This intricate process transforms the Sun's energy into a form that is usable by living organisms, essentially converting light energy into chemical energy. Plants, certain algae, and some bacteria harness sunlight, using its energy to combine water and carbon dioxide into glucose and oxygen.

This photosynthetic pathway not only provides the fundamental source of energy for these organisms but also produces the oxygen necessary for the survival of aerobic life forms, including humans. Our existence relies on this solar-powered synthesis as it stands at the base of the food chain. The glucose produced through photosynthesis serves as fuel for plants and becomes the starting point for the energy flow through ecosystems. The release of oxygen, a byproduct of this process, is vital for us as it replenishes the atmosphere with the very air we breathe. Understanding the dynamics of photosynthesis can enhance our appreciation of the Sun's role beyond just warmth and light. It underscores the interconnection of the solar system and life on our planet. The seemingly simple act of sunlight falling on leaves ignites a series of reactions that underline the balance of our entire biosphere." *SkyKnowledge.org*

Photosynthesis is evidence of a Creator who designed a precise balance, allowing life to be sustained and to thrive on a planet perfectly positioned in relation to its star.

Venus is a wimpy little light as compared to an actual star. It does not give off its own light or sustain life. Was Jesus referring to Venus when He said in Revelation 22:16 "I am the Bright and Morning Star"?

The Real Bright and Morning Star

What is "the Bright and Morning Star"? The brightest, most prominent star? It is our Sun! It is a life-giving light, providing warmth and photosynthesis on planet Earth. The plants, animals, and all mankind cannot survive without it.
 Without the Sun we would all die. Sound familiar? It may seem like a play on words, but without "the Son" we also face death *spiritually* and *eternally,* the very death spoken of in the Garden. Consider the powerful words used in the Scriptures below.

Revelation 21:23 "The city had no need of the *sun* or of the moon to shine in it, for the glory of God illuminated it. The Lamb is its light."

Isaiah 60:19-20 "The *sun* shall no longer be your light by day, Nor for brightness shall the moon give light to you; But the LORD will be to you an everlasting light, And your God your glory. Your *sun* shall no longer go down, Nor shall your moon withdraw itself; For the LORD will be your everlasting light, And the days of your mourning shall be ended."

Psalm 84:11 "For the LORD God *is* a *sun* and shield; The LORD will give grace and glory; No good thing will He withhold From those who walk uprightly."

Malachi 4:2 "But to you who fear My name The *Sun* of Righteousness shall arise with healing in His wings; And you shall go out and grow fat like stall-fed calves."
John 1:19 John the Baptist speaking of Jesus "That was the true Light which gives light to every man coming into the world."

John 8:12 "Then Jesus spoke to them again, saying, "I am the light of the world. He who follows Me shall not walk in darkness, but have the light of life."

The Scriptures above use the Hebrew word **Shemesh** and the Greek word **Helios,** both terms referring to the physical Sun, its rising in the east, and its brightness and brilliance.

As we can clearly see, in the Scriptures above, when Jesus said "I am the Bright and Morning Star" He was not dropping a hint or comparing Himself to Venus (a planet), but to the life sustaining bright and morning star the **Sun.** Jesus is the life-giving Light, the "Light of Life," the very "Light of the World!" *(John 8:12)*

Addressing Alternative Theories

The "Invisible" Star of Bethlehem
The Star of Bethlehem **by Rick Larson**

When I first viewed Rick Larson's film, I loved it—it was captivating and beautifully presented. However, upon much deeper investigation, it became evident that many historical and scriptural pieces of the puzzle were left out. During the time frame it proposes for the birth of Jesus, Herod had already died. Any attempt to claim that Jesus was born or conceived, during a Revelation 12 sign in 3 BC, ignores the fact that this would have occurred after Herod's death, it's been well established by multiple sources as having taken place in 4 BC. But in order to force fit the theory, he claims that Herod died in 1 BC. Not only is this unsupported by ancient historians, but it is also uncorroborated by archaeology, history, or Scripture.

Another important reality is that the sign was **not visible**. The Revelation 12 sign of 3 BC, as depicted in the video, is not accurately represented. In actuality, it occurred **behind the Sun** and was completely obscured by the Sun's glare. Therefore, it could not have been seen by anyone on Earth.

Actual view September 11, 3 BC

In September Virgo is "clothed with the Sun" (Revelation 12:1). This means **it is daytime and cannot be seen;** it is also not the visible constellation in the night sky at that time. This is neither the conception nor the Star of Bethlehem—**it could not be seen, and it occurred after Herod's death.**

Sticking with Scripture, as we should, when referring to the Bethlehem Star, the Magi said, "We saw His star." Furthermore, at the time of Jesus' birth, the Book of Revelation had not yet been written—there was no Revelation 12 sign recorded in Scripture, and there was no New Testament. By the time John wrote the Book of Revelation, Jesus had already lived, died, risen, and ascended into heaven. John was instructed to write down the things that were **yet** to take place in the future. Revelation 4: 1-2 *After these things I looked, and behold a door was standing open in heaven. And the first voice, which I had heard speaking with me like a trumpet, said, "Come up here, and I will show you what must take place after these things."* John wrote down what he saw in the heavens, what was going to take place in the future. Revelation 1:1 *The revelation of Jesus the Messiah, which God gave Him to show His servants the things that must soon take place.*

The sign spoken of in the Book of Revelation 12:1-18 was an astronomical sign that John saw in the heavens from the throne room while he was in heaven. In Revelation 4:1, John wrote down the things to take place "after these things." The Revelation 12 sign shown in the film, cannot be attributed to the date of the birth of Jesus, the Magi said "we saw his star" when it first rose in the east, and it became visible again to them when they left Herod on their way to Bethlehem. *"And when they saw the star, they rejoiced with exceeding great Joy!"* — Matthew 2:10

The 2 BC later Conjunction of Jupiter and Venus.
The June 17th 2 BC date: Jupiter was in conjunction with Venus for a short period in the evening, in the constellation Leo.

Venus does not represent the "King of the Jews" to the Magi. The Magi said, "we saw HIS star" not "Her star".
Venus certainly does not represent a virgin, but the goddess of sex and fertility.

The Magi knew the trajectory and orbit of Venus, it was well known to them, proof of their understanding of the planet Venus, is recorded on Babylonian Tablets housed in the British Museum. Venus was female to every culture at that time: To the Babylonian Hebrews, it was female; to the Babylonians, *Ishtar;* to the Romans and Greeks, *Aphrodite*.

The orbital period of Venus is 224.65 Earth days. With the much slower orbit of Jupiter at 11.86 Earth years, Jupiter and Venus conjunctions occur quite often, roughly every 9 to 12 months. Averaging about 1.04 conjunctions per year over long periods. In just 2000 years, 2000 x 1.04 = **2,080 conjunctions of Jupiter and Venus took place.**

Now let's take a look at how many of those conjunctions took place in the constellation Leo.
The constellations are not the same size; therefore, it can take a longer or shorter period of time for a planet to cross a particular constellation in the planet's orbit (even if they were traveling at the same speed which they are not).
Leo spans 30 deg of the ecliptic (out of 360 deg).
Conjunctions of Jupiter and Venus in Leo
$2,080 \times 1/12^{th} = 173$
Final calculated conjunctions of Jupiter and Venus within the constellation of Leo are **160 to 180 times** in the last 2000 years! By astronomical or any considerations **this was not rare, not unique and in 2 BC Herod was dead!**
Verified by Astronomical programs and numerous historical documents.

Let's remember the Magi and what they said, *"we saw **His** star"* and came to worship the *"King of the Jews"* Not *her* star and not the *Queen of the Jews*.

Conclusion

The above claims for 3 BC and 2 BC are an attempt to force-fit signs that don't meet the first and most important criterion for the Bethlehem Star: **the star must be visible** to align with what the Magi said, "we SAW His star".

Secondly, we cannot simply pick a planet because of a misinterpretation of Revelation 22:16 (bright and morning star) or a misunderstanding of what the Magi believed "His star" to be.

Thirdly you cannot erase history and pretend Herod was not dead. To claim that Herod died in 1 BC, without presenting evidence or by accusing Josephus and other historians of being incorrect is invalid. Not only is a 1 BC date unsupported by ancient historians, but it is also not corroborated by archaeology, history or scriptural events.

We need to set aside our *Christianese* mindset when looking back at 2 and 3 BC.

Scriptural	No	(**not visible**)
Historical	No	(Herod was dead)
Archaeological	No	(no record of a census)
Astronomical	No	(**not visible**)

Another "Invisible" Star of Bethlehem
The Star That Astonished the World by Ernest L. Martin

The author writes that "Herod would have been aware of the outstanding celestial displays that had occurred from May 3 BC to August 2 BC."

The Bible explains it was so unremarkable that Herod had no idea why the Magi were in Jerusalem. The author points to the rising of Jupiter with Venus nearby June 17, 2 BC as part of the sequence. **He admits Venus represents the goddess Ishtar.**

The Actual View June 17, 2 BC

June 17, 2 BC the rising of Jupiter and Venus.
It was invisible.
If it cannot be seen, then it cannot be what the Magi "saw".

The Magi, who said they came to worship "the King of the Jews," would not have followed Venus—the star of Ishtar. Martin goes on to write that, because it was in the constellation Leo, Christians would recognize this as the sign of Jesus, the Lion of Judah. However, there is a problem with this reasoning: it was before Christianity existed, and the Magi were not Christians.

Two obvious problems with his proposal,
1. **Herod was already dead;** his successors had already begun their reigns in 4 BC (2 years earlier) which is historical fact.
2. Jupiter and Venus could not be seen.

The Author claims September 11, 3 BC was the birth of Jesus.

The Actual View September 11, 3 BC

Jupiter joined Regulus September 11, 3 BC. This occurred during the day behind the Sun and **could not be seen.**

Ernest L. Martin's theory is not accepted by the vast majority of historians and biblical scholars.

This is where Martin's historical problem becomes acute. He claims **Luke** made a historical error in compressing or conflating events and was referring to a census whose details we have not yet fully uncovered (no evidence available).
Martin's historical record is incomplete, he speculates there was a unique census in Herod's kingdom that was part of a broader Augustan initiative, which simply hasn't survived in other records (no evidence available).

The academic consensus exists for a reason: the case for Herod's death in 4 BC is built on **multiple, independent lines of evidence that converge on the same date.** To accept Martin's theory, one must believe that the established chronology for the reigns of Herod's sons, the mission of Sabinus, and the associated Roman history are all systematically incorrect.

The Chronological Implausibility: The nativity narrative requires Herod to be alive. Herod died in 4 BC. Therefore, the astronomical events of 3-2 BC are chronologically impossible as the "Star of Bethlehem" that prompted the Magi to visit Herod.

The Census Implausibility: The narrative requires a specific type of empire-wide census that caused Joseph to travel. The only such census to register "all the world's inhabitants" (Luke 2:1), close to Herod's death, is recorded by Augustus in the Res Gestae as having taken place in **8 BC,** a census with **4.233 million** entered into the census roll. Martin's theory requires an unrecorded, local census that would not have compelled Joseph to travel, which contradicts the very reason Luke gives for Jesus being born in Bethlehem.

While Martin's book and his identification of the Jupiter-Venus conjunction is interesting (not rare), it requires the dismissal of a formidable body of historical evidence to make it work. For historians and critical scholars, that is a price too high to pay.

    The weight of evidence is overwhelmingly against Martin's theory. His argument is a classic example of starting with a compelling astronomical idea (the brilliant Jupiter-Venus conjunction) and then trying to force the historical record to fit it, even when that requires overturning a well-established chronology supported by archaeology, Roman history, and the writings of Josephus.

    The entire premise of the "Star of Bethlehem" is that it was an observable phenomenon that the Magi *saw* and interpreted (Matthew 2:2). If the sign that Martin claims occurred on September 11, 3 BC, was completely invisible, then the theory loses its foundation and collapses.

The constellation Leo and Virgo are on the backside of the Sun, behind it and are not visible from anywhere on Earth at this time.

Scriptural	No	**(not visible)**
Historical	No	(Herod was dead)
Archaeological	No	(no record of a census)
Logistically	No	(the nature of the census)
Astronomical	No	**(not visible)**

The sections of his book dealing with the Star read more like sensationalist reporting than a scholarly study; they are filled with speculative argumentation, containing numerous claims that are unsupported or exaggerated. His entire case rests solely on conjecture.

This is why, despite its popularity in some circles, the theory is rejected by historians and astronomers alike.

***The Star of Bethlehem* by Paul Makisker:**
In his video, Makisker mentions a conjunction of Jupiter and Venus on August 12, 3 BC at 3:22 am. Viewed from Tagbuk, Saudi Arabia. He says the Magi came from Arabia because it had been conquered by Babylon, and there are records showing that the last king of Babylon, Nabonidus, wrote that he had gone there due to an evil disease that had plagued him for seven years. According to these records, the one who worships only the Most High God told him he needed to repent, and that this man was Daniel.

Daniel's time, was five and half centuries before the birth of Jesus. The birth of Jesus occurred during the reign of Caesar Augustus from AD 27 to AD 44, at this time Saudi Arabia had long been free from Babylonian rule having collapsed in 539 BC. Makisker is forcing things into place to agree with a predetermined proposed date of the birth of Christ rather than following the evidence. This conjunction was only viewable for an hour and **does not fit the characteristics of "His star" in Matthew.**

The prophecy in Daniel as to when the Messiah would be "cut off", is referring to the crucifixion, yet Makisker places this date when Jesus was Baptized by John the Baptist, counting back 30 years, he then places the birth of Jesus at the 3 BC date, (**after Herod was dead**). He suggests that if Daniel 9 prophesied Jesus' appearance as an adult, then it is reasonable to assume the Magi would have counted back to his birth several decades and would have interpreted the conjunction of 3 BC as a sign of His birth. Not likely, since the Magi saw Venus as Ishtar, feminine and promiscuous, certainly not "His star". He must have been unaware of the Babylonian tablets or Babylonian Talmud. This backtracking he suggests they did, would have been completely unnecessary, as the Magi were exceptional astronomers and understood astronomical events that were about to take place and they would not consider Venus to be "His star".

Makisker's date of August 12, 3 BC

Scriptural	No	(not what Magi described)
Historical	No	(Herod was dead)
Archaeological	No	(no record of a census)
Astronomical	No	(not visible long enough)

Other Astronomical Suggestions and Theories

April 24, 6 BC at 2:15 p.m. Afternoon Saturn and Venus are in conjunction and Jupiter is above.

The Magi would not have considered **Venus** a sign of the Messiah, according to the Babylonian Talmud 156a *"One who was born under the influence of Venus will be a rich and promiscuous person. he will be subject the fire of the evil inclination"*

Jupiter was in Aries and Saturn and Venus were in Pisces, two separate constellations. More importantly, nobody would have seen this conjunction because it was daytime, and it was **below the horizon** before the Sun began to set before 4 pm (way before sunset) and was not viewable, **could not be seen.**

Scriptural	No	(**not visible**)
Archaeological	No	(no record of a census)
Astronomical	No	(**not visible**, Venus = Ishtar)

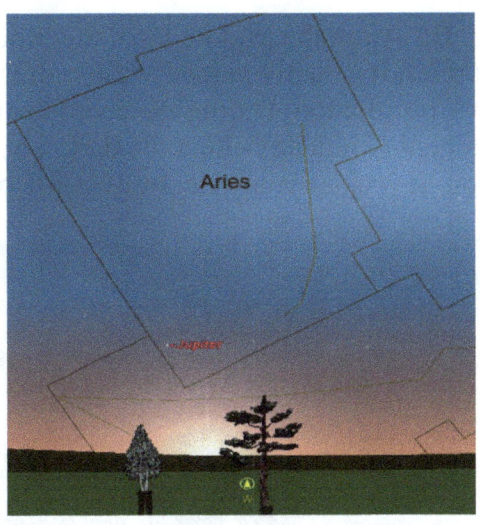

6 BC by M. Molnar

"The Star of Bethlehem: Legacy of the Magi" by Michael Molnar
April 17, 6 BC Jupiter rising in Aries, was invisible from the time it rose in the east, it rose at sunrise and could not be seen. Later while looking into the setting Sun, it washed out and was barely visible for 15 minutes at best. The Magi said "we saw his star", the Magi could not see this, nor could anyone else. An event that was fundamentally unobservable and a very poor candidate.

The Series of Events: Molnar identified a series of astrologically significant events in 6 BC, all involving Jupiter in Aries:

He studied surviving Roman-era astrological texts, according to these texts, the sign for a king's birth, particularly a Jewish king, was associated with the planet **Jupiter** and the constellation **Aries**. He claims Aries was associated with the land of Judea.

A heliacal rising of Jupiter (its first appearance in the dawn sky and being too close to the Sun) in Aries. April 17, 6 BC.

March 20, **6 BC.** This was not a conjunction, but a lunar **occultation** of Jupiter (where the Moon passed in front of it).

On March 20, 6 BC, from the Middle East, the Moon did not just pass *close* to Jupiter; it actually passed *in front* of it, in an event known as a lunar occultation. During the central part of this event, Jupiter was completely hidden behind the dark, non-illuminated side of the Moon. It was, as you state, **impossible to see with the naked eye**.

March 20, 6 BC

Molnar's Defense and the Nuance of His Theory
This is where Molnar's argument shifts from pure astronomy to **ancient astrological interpretation**. His defense would be as follows:

The Significance was Astrological, Not Visual: Molnar argues that the Magi, as master astrologers, would have calculated the positions of the planets using ephemerides (astronomical tables) and other tools. They didn't need to *physically see* the occultation for it to hold profound astrological meaning. The mere celestial fact that the King Planet (Jupiter) was in union with the Moon (a celestial body associated with fertility and, in some systems, the divine feminine) in the constellation of Judea (Aries) was the omen. It was an event that took place in the "heavenly realm," whether visible or not.

We must remember, if it weren't for the Magi, we would not know about the Star of Bethlehem at all. He forgot the Scriptures, which are necessary to understand the astronomical event, and instead interpreted it based on his own idea, disregarding the very witnesses—the Magi—who said, "**We saw His star.**"

The "Star" was a Sequence of Events: Molnar does not pin the entire "Star of Bethlehem" phenomenon on this single, **invisible event.** He presents it as one critical piece in a series of astrological portents throughout 6 BC. The most important of these was the **heliacal** rising of Jupiter in Aries on April 17, 6 BC. (heliacal- invisible)

Another problem with his theory is that **Aries** does not represent Israel, it is **Pisces** instead. Also, **Jupiter** is not the Star of the Jews, it is recognized as a king star, but not "His Star". In the Babylonian Talmud, we see that it is **Saturn** that is recognized as "His Star' By the Jews. (The Magi from Babylon/Persian Empire 539 BC. Babylon/ Parthian Empire 147 BC including 7 BC).

Below are two of the most significant and scholarly criticisms leveled against Molnar's theory. For the Magi, these involve even more important cultural and astrological issues, where his argument becomes much more tenuous.

1. The Constellation: Aries vs. Pisces
Molnar's Claim: Molnar based his association of Aries with Judea on the works of the Roman astrologer Claudius Ptolemy (*Tetrabiblos*), who, writing in the 2nd century AD, assigned Aries to Coele-Syria (which included Judea) and Persia.

The Problem & The Counter-Evidence: This is a weak link for several reasons:

Anachronism and Source: Ptolemy was writing *after* the New Testament period. Applying his specific, Greco-Roman system to Persian Magi over a century earlier is problematic.

The Stronger Association with Pisces: there is a much stronger and more persistent ancient tradition linking the land of Israel with the constellation **Pisces**. This is deeply rooted in Jewish history. The annual reading of the "V'yehi Oros" in the synagogue, which includes a blessing for the rainy season when the sun is in Taurus, also references the "Mazal" (constellation sign of **Pisces (Dagim)** for the month of Adar. This firmly places the Jewish people under the "influence" or "sign" of Pisces in **their own** cultural astronomy/astrology.

The Kepler Connection: This is precisely why **Johannes Kepler's older theory** remains so compelling. He proposed the **triple conjunction of Jupiter and Saturn in Pisces in 7 BC**. The event itself is historically attested in Babylonian astronomical diaries. The conjunction of the two largest planets, happening three times in the constellation symbolically linked to the Jewish people, is a powerful **and *visible*** omen that fits the cultural context far better than a single, invisible event in Aries.

2. The Planet: Jupiter vs. Saturn

The most devastating point against Molnar's core premise.
Molnar's Claim: Jupiter, as the "King Planet" (*Regium Sidus*), was the primary celestial body signaling a royal birth.
The Problem & The Counter-Evidence: While Jupiter was universally seen as a royal and beneficent planet, it was not specifically "the Star of the Jews."

The Talmudic Evidence: The key text is in the **Babylonian Talmud (Shabbat 156a)**, which discusses the influence of one's "Mazal." It states: "*He who is born under [the influence of]* **Shabtay (Saturn)** *will be of a mind whose plans are ruined. Others say: All plans against him will be ruined.*" The Talmud then relates a story where Rabbi Nahman bar Isaac is told by his mother that his star is Saturn (*Shabtay*), and that it is specifically the star of **Israel**.

The Linguistic & Cultural Link: The Hebrew name for Saturn, *Shabtay,* is directly derived from *Shabbat,* the Sabbath day. This creates a powerful and unique theological and linguistic connection between the planet and the **Jewish people**, whose most distinctive religious practice was the observance of the Sabbath. This is not a generic "kingly" association; it is **a specific identifier.**

Implications for the Magi: If the Magi were, as tradition holds, from "the East" (Babylon/Persia), they would have been heirs to the same astronomical and astrological traditions that the Jews of the Babylonian exile had absorbed. They would have been familiar with the seven "wandering stars" (planets) and their associations. It is highly plausible they would have known that **Saturn was the celestial patron associated with the land and people of Israel.**

Conclusion

These two points are not just a minor objection; it strikes at the very heart of Molnar's thesis by demonstrating that his proposed astrological symbolism does not align with the known cultural context of the people the story is about. Together, they successfully dismantle the two foundational pillars of Molnar's astrological argument.

1. He places the sign in the **wrong constellation (Aries)** when the evidence points strongly to **Pisces** as the "Mazal" of Israel. (Mazal ancient meaning "Planet" or "Consetellation").

2. He identifies the **wrong planet (Jupiter)** as the primary signal when the Jewish texts themselves identify **Saturn (Shabtay)** as "their star."

By pointing this out, it shows why the older **Keplerian theory of the Jupiter-Saturn conjunction in Pisces** is a much more culturally and astronomically coherent candidate. It combines the kingly planet (Jupiter) with the planet of Israel

(Saturn), in the constellation of Israel (Pisces), in a series of three spectacularly visible events.

If the Magi said they saw his star, and this specific event could not be seen, then this specific event cannot be "his star" in a literal, visual sense.

Scriptural	No	(couldn't be seen)
Astronomical	No	(couldn't be seen)

The Simpler Explanation: The triple conjunction of Jupiter and Saturn in Pisces (7 BC) points to events that were not only astrologically significant to the Magi but also astronomically correct, spectacularly visible to the naked eye, and aligned with the biblical account.

Conclusion: "His Star"

Was the Bethlehem Star Natural or Supernatural? It was both! All of nature was created by a supernatural God.

The Bethlehem Star was not beyond scientific explanation and it was not a myth. The explanation of the Bethlehem Star is a tangible physical astronomical phenomenon. It was created for a purpose to announce the birth of Christ!

It has been known for centuries, as recorded in the Babylonian Talmud, that the wandering star (planet) Saturn has been associated with Israel. Saturday, named after Saturn, is the seventh day of the week—also *Shabbat*, the day of rest in the commandments given to Moses. According to the Magi **Saturn** was **"His star"**. Unlike the other theories with a conjunction of one planet, then a conjunction with a different planet, or brief alignments with the Moon and Venus that lasted barely an hour, or Jupiter and Regulus that could not be seen, with planets constantly changing partners and shifting between constellations. We see with the triple conjunction in 7 BC, **Saturn** was the same star, along with the same king star Jupiter, in the same constellation throughout the entire event. It was the same star in its rising (*anatole*- as the Magi said) also during the first, second and third conjunction, **consistent** throughout. **"His star"** was consistent from the first to the last conjunction, the beginning and the end. Where have we heard this similarity before?
Hebrews 13:8 *Jesus Christ is the same yesterday, today, and forever.* (NKJV)

The tangible evidence written by Babylonian astronomer-scribes about the triple conjunction in **7 BC** is recorded on their tablets, which still exist today in the British Museum.

To the Magi, the Triple conjunction of 7 BC involving the two **most prophetical wandering stars Jupiter and Saturn,** would have been the rarest astronomical event of their lifetime!

The Triple Conjunction of 7 BC

Scriptural   check
It fits the description by the Magi, it rises in the east and sets in the west.

Historical   check
Herod was **alive,** this occurred before Herod's successors began their reigns in 4 BC. Before the Lunar Eclipse, and the historical Golden Eagle Event.

Archaeological   check
It was during the census of Caesar Augustus in 8 BC
Res Gestae Divi Augusti
Recorded on Babylonian astronomical tablets located in the British Museum.

Astronomical   check
Bright and visible from its rising *anatole*. All three conjunctions occurred in the same constellation Pisces by the same planets (wandering stars) Jupiter and Saturn.

Shabtai

Saturn, being consistent throughout the
Triple conjunction of 7 BC,
there is no more fitting candidate and representation for

"His star"

Appendix

Calendars and Astronomical Dates

Calendars serve as a measuring stick, whether Gregorian, Julian, or Hebrew. By studying astronomy—God's own calendar—we can observe events in the heavens that fulfill biblical prophecy and align with scriptural descriptions and timelines. The Gregorian Calendar provides a convenient reference point to locate these astronomical events and determine when they occurred in the sky. Today, we use the Gregorian Calendar in astronomy programs, such as Starry Night, because it is the calendar most familiar to us. These dates can also be converted to the Julian Calendar if needed.

For example, the Triple Conjunction occurred with its first alignment on May 29, 7 BC, the second on September 27, 7 BC, and the final on December 20, 7 BC. **These dates are plotted on the Gregorian Calendar,** but it is far more important to understand the event in light of Scripture first. We should not begin with a predetermined date for Christ's birth or crucifixion based solely on videos, hearsay, or tradition alone. Notice I emphasize *tradition alone*. While tradition can be helpful, it is not always reliable and must be supported by additional evidence—archaeological, historical, astronomical, and even biological—that aligns with Scripture.

The guiding principle is Scripture first, supported by evidence to determine dates. Thanks to advances in astronomy and archaeology, we can now identify biblical events and their timing with far greater accuracy than in the past.

Julian Calendar vs Gregorian Conjunction Dates A critical point that is key to resolving dating issues for the conjunctions. The discrepancy is almost entirely due to the difference in these calendars. The astronomically correct Gregorian calendar in the software program Starry Night versus the dates often cited in historical texts are based on the Julian calendar in use in 7 BC.

By the year 7 BC, the Julian calendar had already accumulated an error against the tropical (solar) year. The Gregorian calendar was introduced in 1582 to correct this drift. The key is:

The discrepancy was much smaller in the 1st century and grew significantly larger by the 1500's. The gap was about 1 day by the 1st century and grew to 10 days by the 1500's.

Detailed Explanation

The problem stems from the fact that a solar year (the time it takes the Earth to orbit the Sun) is not a whole number of days. It is approximately 365.2422 days.

Julian Calendar (Introduced in 45 BC)

Rule: Decreed by Julius Caesar, this calendar had a simple rule: a year is 365 days long, with a leap year of 366 days every four years without exception.
Length:This makes the average Julian year 365.25 days.
The Error: This is 0.0078 days (about 11 minutes and 14 seconds) longer than the actual solar year.
The Result: This tiny overestimate causes the calendar to slowly drift forward relative to the actual seasons. The equinoxes and solstices occur slightly earlier each year in the calendar.

The Discrepancy Over Time
Tracking the cumulative error:

By the 1st Century (c. 100 AD)
About 125 years had passed since the calendar's introduction. Error Accumulated: 125 x 0.0078 days/year = 1 day. So, a celestial event like the vernal equinox would have been occurring around March 22 on the Julian calendar instead of its original date of March 25.
By the 1500's (specifically by 1582): About 1627 years had passed since 45 BC.
Error Accumulated:1627 years x 0.0078 days/year =12.7 days.
In practical terms, the vernal equinox, which is crucial for calculating Easter, was falling around March 11 instead of its

intended date of March 21. This 10-day error was the immediate problem the Catholic Church needed to fix.

The Gregorian Reform (1582)

To correct and prevent the 10-day drift from happening again, Pope Gregory XlII introduced the Gregorian Calendar with two key changes:

The One-Time Correction: To bring the equinox back to March 21, 10 days were skipped. Thursday, October 4, 1582, was followed by Friday October 15, 1582.

The New Leap Year Rule: The rule of " a leap year every 4 years" was refined. Now, a year is a leap year if: It is divisible by 4, except if it is divisible by 100, unless it is also divisible by 400. This means 1700, 1800, and 1900 were not leap years, but the year 2000 was.

Length: This makes the average Gregorian year **365.2425 days,** which is extremely close to the solar yea of 365.2422 days.

The "Astronomical" or "Historical Alignment" Method

(Most Common Answer) This is the method used by historians and astronomers. The key is to work backwards from 1582, when the 10-day correction was made.

The Rule: The Julian calendar was running slow, so Julian dates are later than the astronomical reality. To convert the Julian date to the Gregorian one, you add days.

The Math for 7 BC:

The error accumulates at a rate of about 0.0078 days per year. From 7 BC to AD 1582 is 1,588 years. 1,588 years x 0.0078 days/year = 12.4 days

The Calculation: We take the proleptic Gregorian date (the "real" astronomical date) and find its Julian equivalent. If the Gregorian date was September 25, 7 BC, the contemporary Julian date would have been Gregorian date + 12 days = September 27, 7 BC.

This 2-difference is much smaller than the 10-day difference in 1582 because the error had 1,588 fewer years to accumulate.

Are we having fun yet?

So, when discussing this event, it's important to know the date recorded in historical (Julian) terms corresponds to an earlier date on our modern (Gregorian)calendar.

Astronomical Records: Programs like Starry Night and professional astronomers always use the Gregorian calendar (proleptically, for dates before 1582) because it accurately reflects celestial positions. So, the event is permanently and officially cataloged as September 27, 7 BC for the second conjunction (the proposed Birth of Jesus).

When we project our modern Gregorian calendar backwards, we are applying a correction.

Johannes Kepler was using the Julian Calendar, which by his time (early 1600's) was 10 days out of sync with the seasons.

Kepler lived from 1571 to 1630. He did his calculations on the Star of Bethlehem before the Gregorian reform was widely adopted in Protestant regions like Germany. He was using the Julian calendar.

So, when Kepler identified the conjunctions as occurring on October 6 and December 30 (Julian), he was absolutely correct for the calendar he was using. Modern astronomy software like Starry Night simply recalibrates those same astronomical events back onto the more accurate Gregorian timeline, revealing the true dates of September 27, 7 BC for the second conjunction and December 20, 7 BC for the third and final conjunction.

There are two main types of conjunctions;

Heliocentric (Geometric): This is when two planets are at the same ecliptic longitude as seen from the Sun. It's a single, precise moment in time.

Geocentric (Apparent): This is when two planets appear closest to gether in the sky as seen from Earth. This is what ancient astronomers would actually observe. Due to planetary motion and parallax, this "closest approach" can happen on a different date than the heliocentric conjunction. This difference explains the gap in dates.

The discrepancy between the dates given by Johannes Kepler and Dr. David Hughes, astronomy professor, is not due to the use of the Julian versus the Gregorian calendar—both modern scholars use the Gregorian for calculation. The difference arises because:

Kepler was referring to the **observable, geocentric climax** (c. December 30).

Hughes was referring to the **theoretical, Heliocentric conjunction** (c. December 1).

Therefore, a date that was **October 6 on the faulty Julian calendar** in 7 BC would be **September 27, 7 BC on our modern, corrected Gregorian calendar**.

The **October 6** date is the Julian equivalent of Starry Night's **September 27 (Gregorian)** date.

The **December 30** date is the Julian equivalent of Starry Night's **December 20 (Gregorian)** date.

Dates from Starry Night are **astronomically more precise** for a modern audience because they are calibrated to our current calendar and the actual solar position. When we talk about the "sky as seen by the Magi," the planetary positions you see in Starry Night on **May 29, September 27** and **December 20, 7 BC (Gregorian)** are **exactly what they would have witnessed.**

The historical dates (late May, Oct, Dec) are still useful for understanding what Kepler calculated and for reading older scholarly work, but for visualizing the event, the Gregorian dates are correct.

Star of Jacob
Numbers 24:17
"I see Him, but not now; I behold Him, but not near; A Star shall come out of Jacob; A Scepter shall rise out of Israel, and batter the brow of Moab, and destroy all the sons of tumult. (NKJV)

Balaam had a vision of Jesus in his head, he said, "I see HIM" and "I behold HIM". His words must be understood as having reference to One whom he beheld with the eyes of his mind. This Star is referring to a brilliant personage. *A Scepter shall rise out of Israel.* This further defines **this** particular "star ' in the above Scripture as a ruler of men.
He prophesied of a Star and a Scepter that would arise in Israel. He foretold of a great King who would come forth from Jacob, one who would conquer Moab and all their enemies. Christ is the One who will fully and finally bring an end to the raging of the nations.

Why is Matthew the only one to record Herod's the murder of the innocents?

Professor William F. Albright estimates the population of Bethlehem at the time of Jesus' birth to be about 200- 300 people. The number of male children, two years old or younger, would be about 6 or 7. Other scholars claim the number was between 10 to 20 male children in Bethlehem and the surrounding area. Although their deaths were a terrible atrocity, it would have been comparably very low to other events at that time and they were Jewish babies and may not have been considered noteworthy to a Roman writer. However, it was more likely Herod would not have wanted it to be known that he had killed toddlers, he wanted it kept secret since in his warped way, he still was trying to appease the Jews. He specifically wanted the toddler boys, those in Bethlehem eliminated and nowhere else in Israel and had it done at night or very early morning before sunrise. We know according to Scripture that Joseph took Mary and Jesus and fled in the night.

Was there enough time between the Lunar Eclipse of 4 BC and Passover?
Does the narrative of Josephus' events allow enough time between the death of Herod and the following Passover in 4 BC?

By considering the evidence it is made clear there was plenty of time

After the Golden Eagle event, during the eclipse of March 13, 4 BC Herod was already sick and dying "on his couch" and could only raise himself on his elbows. It was due to the young Jewish rebels believing Herod had already died, that they became emboldened to take down the Golden Eagle Herod had placed above the temple gate. After tearing it down, Herod had the perpetrators executed and that night was the Lunar Eclipse March 13, 6 BC. Passover was one month away. At the direction of his physician Herod "bathed himself in the warm baths that were at Callirrhoe"(only 6 miles from Jericho) and realized there was no hope of recovery. Herod attempted suicide, he was in much pain and near to death. He asked for an apple and attempted suicide with the knife, but his cousin stopped him and he let out a loud cry. Antipater II (his son) was being held in captivity, heard about it, was joyous since he hated his father and thought he could now take the throne. The jail guard told Herod how Antipater had responded thinking Herod had died, so Herod wasted no time in having him slain. He had previously obtained the permission from Augustus to slay Antipater if he chose to do so. Five days later Herod died by Passover April 11, 6 BC. Herod was dead, Archelaus mourned only seven days. *Josephus Antiquities of the Jews Book XVII Chapters 5-9*

The events below happened after Passover.

Herod's sister "Salome I" and family had gone to Rome to let Caesar Augustus know Archelaus should not be king, at the same time Archelaus had arrived there. Sabinus who Archelaus had put in charge in Jerusalem was causing riots. Augustus therefore did not give Archelaus kingship but instead Ethnarch of Judea, along with his brother Herod Antipas and Phillip who became Tetrarchs and Herod's sister "Salome I" received her territories. This way Augustus remained King over all.

"Now Antipater IV, Salome's son, a very subtle orator, and a bitter enemy to Archelaus (Herod's other son), spake first to this purpose: That it was ridiculous in Archelaus to plead now to have the kingdom given him, since he had, in reality, taken already the power over it to himself, before Caesar had granted it to him; and appealed to those bold actions of his, in destroying so many at the Jewish festival; and if the men had acted unjustly, it was but fit the punishing of them should have been reserved to those that were out of the country, but had the power to punish them, and not been executed by a man that, if he pretended to be a king, he did an injury to Caesar, by usurping that authority before it was determined for him by Caesar; but if he owned himself to be a private person, his case was much worse, since he who was putting in for the kingdom could by no means expect to have that power granted him, of which he had already deprived Caesar [by taking it to himself]
(Caesar Augustus) "he made no full determination about him; and when the assembly was broken up, he considered by himself whether he should confirm the kingdom to Archelaus, or whether he should part it among all Herod's posterity; and this because they all stood in need of much assistance to support them."
Josephus Antiquities Book 17 chapters 5-9
Caesar Augustus did part the kingdom to Herod's children in 4 BC, therefore maintaining his own control as Emperor.
"for he (Herod) *was about the seventieth year of his age"*, during the golden eagle event, he died only weeks later, he was born in 74 BC, therefore he died in 4 BC.

Additional Evidence Confirming Herod's 4 BC Date of Death

SABINUS (end of first century BC), Roman official. Sabinus, then Augustus' treasurer in Syria, was sent to Judea after Herod's death in 4 BC to take charge of the latter's estate as procurator. On his arrival he acceded to the request of Varus, governor of Syria, to hand over the custody of the citadels and treasures to Herod's son Archelaus, pending Caesar's decision concerning Herod's will. However, immediately after the departure of Varus and Archelaus for Antioch and Rome, respectively, he took possession of the royal palace and demanded from the custodians particulars regarding Herod's treasure. Sabinus' conduct caused a revolt on the festival of **Shavuot**(Pentecost 50 days later), when many pilgrims had assembled in Jerusalem. Sabinus seized the Tower of Phasael, from which he gave the signal to attack the rebels. As the battle developed, the Romans set fire to the Temple chambers, capturing and plundering the Temple treasury. These acts further enraged the people, and they besieged the royal palace where Sabinus and his followers had fortified themselves. The Jews demanded that the Romans leave the city, offering to spare their lives, but Sabinus would not trust them. Riots continued throughout Judea until Varus hurried back to suppress them. When he reached Jerusalem, Sabinus fled. BIBLIOGRAPHY:

Jos., Ant., 17:221–94; Jos., Wars, 2:16–74; Schuerer, Hist, 161f.; Klausner, Bayit Sheni, 4 (19502), 173–7; Pauly-Wissowa, 2nd series. 2 (1920),

Herod's death date of 4BC is not solely based on Josephus alone.

Mark Antony, Octavian (later named Caesar Augustus), and the **Roman Senate appointed Herod the Great as King of Judea** after the Treaty of Brundisium sometime in early to **mid-October of 40 BC,** and within 7 days thereafter Herod sailed back to Ptolemais (Acco) on the coast of Judea. This would be a short 10-15 day leeward (downwind) voyage arriving in late October or early November of 40 BC, in time for Herod to gather an army

and begin a campaign against Antigonus and the Parthians (per Josephus Ant. 14.14.4-14.15.1 and Wars 1.14.4-1.15.4), ultimately capturing Antigonus and Jerusalem in 37 BC, Herod's 3rd year since being made King (by Tishri reckoning).
Analysis of Herod the Great's 40 BC voyage to Rome and return http://theos-sphragis.info/herods_40bc_analysis.html
The Greek historian, Appian, the Perusine war Appian's account in The Civil Wars book 5 1.10-6.52,
The Pact or Treaty of Brundisium Sept/Oct 40 BC
The Concordia Casinum Inscription and Virgil's 4th "Eclogue"

Herod was born in 74 BC. He was in his seventieth year when he died during the Golden Eagle Event and Lunar eclipse, which means he died in 4 BC.

The only way to conclude that Herod died in 1 BC or any other year than **4 BC is to ignore all the evidence.**

Additional Confirmations for The Birth of Jesus.
Josephus informs us that Herod died between a Lunar eclipse and shortly before a Passover (*Antiquities* 17.9.3, *The Jewish War* 2.1.3), based on the previous evidence making a lunar eclipse in March 4 BC the year. Josephus writes that Herod reigned for 37 years from the time of his appointment in 40 BC and 34 years from his conquest of Jerusalem in 37 BC (*Antiquities* 17.8.1, *War* 1.33.8). Using so-called inclusive counting, this too, places Herod's death in 4 BC.

According to Origen and Eusebius the Holy family returned from Egypt in the **first year** of the reign of (Herod's son) Archelaus, who began his reign in April of 4 BC. Joseph was told in a dream the one who sought to kill the child (Herod) was dead. If Herod had died in 1 BC as a few claim, the family would have returned when Herod was still alive. They returned in the first year of Archelaus' reign which began in 4 BC, after Herod was dead.

Additional Babylonian Information
In the Babylonian system in 1500 BC Jupiter, the largest and brightest planet, was known as the star of Marduk, the supreme god of Babylon. **Saturn,** the second largest planet, **was the star of the king,** the earthly representative of the god. The worship of Marduk was prevalent in 1500 BC well before the Captivity of the Jews in Babylon and before the Magi in the gospel of Matthew. (See more in the chapter "The Magi the Wise Men of Babylon")

Jewish Calendar Days
In the Bible when God created time, He first created night and then day. A Jewish calendar day begins with the night beforehand. While a day in the secular calendar begins and ends at midnight, a Jewish day goes from sunset to sunset. Shabbat begins on Friday night a few minutes before sunset.

"Saturn and the Rainbow: Created by God and Belonging to Him"

Genesis tells us that after the flood God placed a rainbow in the sky. Imagine the awe Noah felt looking into the sky and seeing a beautiful rainbow for the first time! It was a sign in the sky, the sign of the promise that God would never flood the whole Earth again. Since then, most people have seen rainbows multiple times throughout their lives, we take pictures of them and wonder where does the rainbow end, and thoughts about Leprechauns and twinkling pots of gold.

The legend of Leprechauns is one of the most famous stories in Ireland. It refers to a magical kind of fairy, that are members of the Tuatha De Danann, a race of supernatural beings in ancient Irish mythology, who invaded Ireland and were banished to live in the other world known as Tir na nog.

I will not go into anymore depth about leprechauns and fairies, but it is clear this is just one of the many ways the true meaning of the rainbow has been covered up, redirecting our thoughts to think of pots of gold, rather than our attention being drawn to God and His creation and promises.

A similar situation is the case of the planet Saturn.

There has been information circulating about the ancient Romans and what they believed about Saturn, as well as information about Roman Catholics and their reason for celebrating Christmas on December 25th. The reason given is that it was the solstice and the celebration of Saturnalia, suggesting that Saturn is the planet of a false god—or even Satan. This is not surprising; of course, Satan is going to hijack Saturn in an attempt to create confusion and fear in those who watch and desire to understand God's creation. The normal tactic of an enemy is to corrupt a message.

In this book, we are committed to Scripture and what the Magi believed, what the Babylonian Talmud may have taught them during the captivity of the Jews in Babylon, and ultimately motivated them to seek, find, and worship the King of the Jews. Not what the Romans or the Catholic church later believed; Catholicism did not exist in 7 BC, or before Christ had risen and ascended.

The winter solstice is due to the Earth's orbit and tilt, a planetary cycle, God created it and called it good. Saturn does not belong to Satan or any other false god, it was created by God and for God to be used for His purpose and plans. Let's remember the Lord owns every planet, including Saturn and Jupiter, and all the stars, and all the cattle on a thousand hills.

Psalm 50:10 — "For all the animals of the forest are mine, and I own the cattle on a thousand hills." In this verse, God is speaking —reminding the people that He is the Creator. He made everything, even the cattle on a thousand hills. God created each planet, comet, and star with specific detail. Their paths, orbits, and trajectories were also designed by Him. Every part of His creation is purposeful, thoughtfully made, and loved by Him.

Notes

Astronomy Terminology

New Moon: Technically a new moon is completely dark. The Moon is in between the Sun and the Earth and the backside of the Moon is illuminated by the Sun. As a result the Moons dark side is facing the Earth making it invisible in the night sky.

Waxing Crescent: The Waxing Crescent phase of the Moon follows the New Moon, signifying the beginning of the Moon's visibility from earth. During this phase, a slim, crescent-shaped sliver of the Moon is illuminated by the Sun and becomes visible against the night sky. This crescent gradually grows in size over several days.

Full Moon: The Full Moon phase is a striking and well-known stage in the lunar cycle, characterized by the entire face of the Moon being illuminated by the Sun. This occurs when the earth is positioned directly between the Sun and the Moon, allowing sunlight to fully light up the Moon's surface.

The origins of the word "planet" comes from the Greek word meaning "wanderer"

Who were they?

Gaius **Suetonius** Tranquillus AD 69 – AD 122
The Roman historian Seutonius about 120 AD
Roman Historian and Biographer
 "An old and well established belief was held all over the orient, that one would arise from Judea who would establish a government over all men." (The lives of the Caesars -Life of Vespasian 4.5)

Eusebius of Caesaria AD 260/265 – AD 339
"Father of church history"
Eusebius was a scholar of the biblical canon.

Origen of Alexandria AD 185 – AD 253
Early Christian Scholar and Theologian

Josephus Flavius Josephus was a first-century Jewish historian personally involved in the *Great Jewish Revolt against Rome*. A former political leader and priest, he is our most important witness to Jewish life and history at the close of the biblical period (first century 36- 100 CE).

Astronomical Diaries and Related Texts from Babylonia
By the late Abraham J. Sachs
Completed and Edited by Hermann Hunger
Volume I Diaries from 652 B.C. to 262 B.C.
Volume II Diaries from 261 B.C. to 165 B.C.
Volume III Diaries from 164 B.C. to 60 B.C. Verlag der Osterreichischen Akademie der Wissenschaften Wien 1988

Astronomical Diaries and Related Texts from Babylonia with notes (spirasolaris.ca)

DIARIES 1 The texts edited here are usually called "diaries" or "astronomical diaries" by modern authors. The Akkadian term for them is na ru ša ginê "regular watching" which is written at the end and on the edges of the tablets. That a regular watch was kept by observers specifically trained and employed for this purpose is shown by two documents dealing with such employment by the assembly of the temple Esangila in Babylon . 2 In these documents the term na ru na ru is used for one of the duties of the employees, and it seems very likely that this can be translated "to make regular observations". They also have to "give" to their employer yearly tersêtu and meš.hi . ters tu occurs in the colophons of astronomical tablets (cf. ACT p.22f.) where it seems to refer to the tablets mes themselves or their contents; meš.hi is the word for the texts called "almanacs" by A. Sachs in JCS 2 277ff. It is therefore likely that the same people who had to make observations also prepared almanacs and astronomical tables. Diaries were filled with entries day by day as the observations were made. This can be seen from the "short diaries" which cover from a few days to a little over a month.

The earliest diary found so far concerns the year -651. We know however that observation of the sky with the intention to control the observed phenomena is older than this date. Eclipse reports preserved on tablets go back to the second half of the 8th century BC.

Since the Babylonian day began with sunset, the diaries record first the events of the nighttime, and then of the daytime segment of a day. Nighttime is identified as such by the word GE "night" before the day number; daytime has no specification. For a given night, the weather situation is usually reported first; then follow lunar and finally planetary observations. During daytime, weather phenomena naturally dominate.

The year is not defined by some independent observation of natural phenomena. It is intended to have the same seasonal events occur at approximately the same point in the year; on the other hand, the year has to contain a whole number of months. The number of days in a year is of no interest in the Babylonian calendar. Since twelve months are about 11 days too short to make up a full solar year, an additional month has to be added to about every third year to keep the seasonal events at the same place in the year. This custom of adding a month to a year can be found in the oldest documents which give any calendaric information (from the third millennium) and continues to the end of the attested use of cuneiform writing. In the first millennium B.C., there are two points in the year where an intercalary month can be added: either after the sixth or after the twelfth month. Until about the middle of the first millennium B.C., this intercalation was done when one felt that it was needed. We know this from royal letters commanding intercalation as late as the reign of Nabonidus and from other official letters on the same topic from the time of Cyrus or even Cambyses. From -380 onward, a fixed pattern of intercalation was followed which had seven intercalary years in a 19-year period. Even before this time, one had evidently tried to establish such a pattern, as can be seen from the ahnost but not quite regular distribution of intercalary months during the preceding centuries. In the astronomical compendium mul APIN, several rules for deriving intercalculation from the observation of stars or the moon can be found. It is uncertain to what extent these rules were utilized; there is no reference to them in the scholarly letters of the 7th century royal correspondence from Nineveh (edited by S. Parpola in LAS). The Babylonian year begins in spring, around the vernal equinox; obviously, true lunar months do not permit the beginning of the year to remain at a fixed distance from the equinox. The relation between the beginning of the year and the equinox can be followed with the help of the tables in PD; computations were done by F. X. Kugler, SSB II 435ff. and Erg. 227ff. The Babylonian day begins with sunset. In the diaries, the night is divided into three parts which correspond to the three watches of the night, as can be seen from the abbreviated terminology. These technical terms are: USAN "first part of the night" MURUB ZALÁG 4 "middle part of the night" "last part of the night"

Note as far back as 600 BC to 0 the names of the "normal stars" or "predictable stars" are the brightest star within a constellation.

The following table lists the Babylonian names, their translation, the modern names, and the ecliptic coordinates for the years -600, -300, and 0, of the usual Normal Stars:

Pisces MÚL KUR šá DUR nu-nu The bright star of the Ribbon of the Fishes Piscium 350.73/5.23 354.87/5.24 359.02/5.26
Aries MÚL IGI šá SAG HUN The front star of the head of the Hired Man Arietis 357.88/8.39 2.02/8.40 6.17/8.41
MÚL ár šá SAG HUN The rear star of the head of the Hired Man Arietis 1.52/9.90 5.67/9.90 9.82/9.91

The names Pisces and Aries.

Pisces was the two fish connected by a ribbon.

Aries was not viewed as a Ram but a Hired Man.

Note: When Solomon died, between 926 and 922 BC, the ten northern tribes refused to submit to his son, Rehoboam, and revolted. From this point on, there would be two kingdoms of Hebrews: in the north - Israel, and in the south - Judah. The Israelites formed their capital in the city of Samaria, and the Judaeans kept their capital in Jerusalem. The kingdom split in two, approximately 300 years before the Babylonian captivity and the two fish in Pisces has been considered to represent the two tribes of Israel.

Jeremiah 50:32-34 The arrogant one will stumble and fall with no one to pick him up. And I will kindle a fire in his cities to consume all those around him. This is what the LORD of Hosts says: "The sons of Israel are oppressed, and the sons of Judah as well. All their captors hold them fast, refusing to release them. Their Redeemer is strong; the LORD of Hosts is His name. He will fervently plead their case so that He may bring rest to the earth, but turmoil to those who live in Babylon.

Sources

Jason Schlude Assistant Professor of Classics at College of Saint Benedict and Saint John's University Specializes in the relationships shared by ancient Rome, the Near East and the Parthian empire. Associate Director of the archaeological excavations at Omrit in northern Israel.
The Shiloh Excavations
When did Herod the Great die? Part 1
When was the Lunar Eclipse?
Rick Lanser MDiv

Flavius Josephus Antiquities of the Jews Book XVII
From the Death of Alexander and Aristobulus (Herod's Death) to the Banishment of Archelaus

Perseus Digital Library - Antiquities of the Jews
Flavius Josephus, Antiquities of the Jews, Book 1, Whiston chapter pr. (tufts.edu)

Date of Herod's Death
The Year Herod Conquered Jerusalem Part lll of Sabbath Year of 36/37 BC
Historical writings of Josephus, Pliny, Philo
https://www.yahweh.org/publications/sjc/sj18Chap.pdf

Bible Encyclopedias
The 1901 Jewish Encyclopedia   Sabinus
Singer, Isidore, Ph.D, Projector and Managing Editor. Entry for 'Sabinus'. 1901 The Jewish Encyclopedia.
https://www.studylight.org/encyclopedias/eng/tje/s/sabinus.html. 1901.

Associates for Biblical Research
The Parthinian War Paradigm and The Reign of Herod the Great
Rick Lanser MDiv

We have looked at many objections critics propose in attempts to discredit the death date of Herod in 4 BC. The article below offers an in depth study covering every aspect of the objections, while supporting with historical facts details pertaining to Herod's death.

Associates for Biblical Research
When did Herod the Great die? Part 1
Rick Lanser Mdiv May 15, 2019

The Intrigues of Salome I, Herod the Great's Sister - Marg Mowczko

Understanding Roman Currency: *A Comprehensive Guide to Ancient Coin The Roman Empire* (roman-empire.net)

Bible Archaeology Report: *Caesar Augustus: An Archaeological Biography* – Bible Archaeology Report
Author Bryan Windle
www.Biblearchaeologyreport.com

The Deeds of the Divine Augustus
By Augustus Written 14 A.C.E.
Translated by Thomas Bushnell, BSG
The Internet Classics Archive | The Deeds of the Divine Augustus by Augustus (mit.edu)
Copyright 1998, Thomas Bushnell, BSG. This translation may be freely distributed, provided the copyright notice and this permission notice are retained on all copies.

American Numismatic Society
OCRE Online Coins of the Roman Empire Online Coins of the Roman Empire (numismatics.org)

Antioch in the Orontes
Moneta Historical Research by Tom Schroer
Ancient Roman coins from Antioch in the Forum Ancient Coins consignment shop.

The Babylonian Captivity and Its Consequences
By Robert Drews Professor of Classical Studies Emeritus at Vanderbilt University.
Specializing in ancient history, prehistory, and the evolution of warfare and religion

Bible Verse Study.com
Quirinius – Proof The First Census of Quirinius Was In 8-4 BC
Biblehub.com
Quirinius Governor of Syria

Biblical Archaeology following the Babylonian Captivity
Israel restored after the Babylonian captivity (truthnet.org)

The Mishna Project
Shekalim Chapter 4 Mishnah 7
Sailing Seasons in the Mediterranean in Early Antiquity,"
Mediterranean History Journal 20: 145–162,

Source Abarim Publications' free online interlinear (Greek/English) New Testament, translated word by word and with Greek grammar parsing codes

Encyclopedia of The Bible – Host of Heaven
BibleGateway
Bibliography C. H. Gordon, *Ugaritic Literature* (1949); S. N. Kramer, ed., *Mythologies of the Ancient World* (1961); T. H. Gaster, "Sun," "Moon," IDB (1962); J. Gray, "Ishtar," "Shahar," IDB (1962); M. Tsevat, "Studies in the Book of Samuel," HUCA (1965), 49-58.

A Guide to Understanding the Phases of the Moon
(thenightsky.com)

Cometography.com
Gary W. Kronk's Cometography
21P/Giacobini-Zinner *(cometography.com)*

Jupiter Family Comets Interactive
There are 786 Jupiter Family Comets in this database
Jupiter-family Comets Space Reference

EarthSky Giacobini-Zinner 21/P
Best pics of the comet as it swept safely past | Space | EarthSky

Astronomy Professor David W Hughes
Department of Physics and Astronomy
University of Sheffield

Away in a Manger, But Not in a Barn:
An Archaeological Look at the Nativity (biblearchaeology.org)
Greg Byers MA The Life & Ministry of the Lord Jesus Christ & the Apostles 26-29 AD

Israel Climate Weather By Month, Average Temperature
Weather Spark

Dr. William A. Simmon Professor of New Testament and Greek at Lee University Scholar in the field of Bible theology.

The New Testament Calendar
by Ernest L. Martin, Ph.D., 1996
Associates for Scriptural Knowledge
Associates for Biblical Research
Did Ezra Come to Jerusalem in 457 BC? (biblearchaeology.org)
17 Jewish Calendar Facts
by Yehuda Altein

Hebrew Calendar (cgsf.org) This is a calculated calendar, it is possible the date proposed for the Passover and other feasts can be off a day or two since it is the *visible* moon that actually determined the dates of the festivals in the biblical period.

Rosh Hashanah: How the Jewish New Year Connects to Jesus
An Invitation to Who You Already Are by Bible Project Scholar Team Sept 7, 2023 *Rosh Hashanah: How the Jewish New Year Connects to Jesus (bibleproject.com)*

Mystery of the Christmas Star MSUM Planetarium
An Astronomers Confirmation by Dr David Hughes
BBC The Sky at Night

Chris Walker curator at the British Museum

NASA Jet Propulsion Laboratories
California Institute of Technology
Center for Near Earth Object Studies CNEOS

Bible Archaeology: *Israel Restored after the Babylonian Captivity (truthnet.org)*

Forgotten Shavuot History: *The 4 B.C.E. Rebellion and the Therapeutae – TheTorah.com*
Prof. Martin Goodman is Emeritus Professor of Jewish Studies at the University of Oxford, Emeritus Fellow of Wolfson College, and Supernumerary Fellow of the Oxford Centre for Hebrew and Jewish Studies. He holds M.A., D.Phil. and D.Litt. degrees from the University of Oxford.

About the Author

Seeking the Truth

I was born and raised in a large, very Catholic family, and all of us children attended Catholic school in our early years before continuing our education in public school.

During grade school, I participated in "Released Time" trailers for religious instruction, which were portable classrooms. One was designated for Catholic students, and another for Protestant students. Out of curiosity, and with a touch of fear of excommunication, I quietly slipped into the line for the Protestant trailer, and in that trailer I first heard about the love of Jesus. That experience inspired me to begin reading my Bible.

At the age of twelve, I was about to go through Confirmation. I felt that if I was to be a Catholic, I wanted to be a good one, and I needed to have my questions answered. On Sunday morning, following Catholic Mass, I carried my Bible into the parish hall and asked the priest, "It says in the Bible we are not to bow down to any statues, but they are all over the inside of the church. Why do we kneel in front of them?" The priest replied, "How dare you question me, you rebellious child! You are not allowed to read the Bible—only priests are allowed to read the Bible!" That moment sent me on a **quest for truth.** It took a couple of years, but I found it. — John 14:6-9

In 1975, at Calvary Chapel Costa Mesa, I became a Christian. I had my own plans, graduated early from high school and began college during my senior year. After ten years of working in the medical field, God changed my plans and my path.

The Author

From those early experiences, A.C. Katz's walk with Christ unfolded, leading to a ministry that touched many areas of Christian service and communication. A.C. became actively involved in Christian Disability Ministries and Christian music ministry, serving as a recording artist with Maranatha Records. A.C. was also a live radio host on KCLB/KLOVE Christian Music Radio during its formative years and expansion in the 1980s, holds an RMM in radio, and works as a voiceover artist and copywriter for both Christian and secular media.

An amateur astronomer, **A.C.** has appeared as a guest on *Prophecy Watchers* with Mondo Gonzales, *Jimmy Evans' Tipping Point*, and *Janie DuVall's Last Days Encounters & End Time Prophecy*. The author has studied Bible prophecy under the teachings of respected scholars and ministries, including Chuck Missler, Dr. David Reagan of Lamb & Lion Ministries, Grant Jeffrey, Drs. Jack and Rexella Van Impe, and Dr. Gary Stearman of *Prophecy Watchers*.

A.C. Katz is the author of *The Great American Writing on the Wall and From Babylon to Bethlehem and Back Again: The Bethlehem Star Revealed and Revelation 12: The Sign—The Point of No Return.* In the latter book, the author explores the history and profound significance of two of the most extraordinary astronomical events mentioned in the Bible, the Bethlehem Star and the Revelation 12 Sign. Drawing upon theological, historical, archaeological, and astronomical evidence, Katz demonstrates how God used these celestial events to introduce His Son to the world and to reveal end-time events.

"My desire is to strengthen people's faith, by revealing the hand of God in creation, recognizing His majesty, and to give Him the glory and honor that belong to Him alone." A.C. Katz

www.ingramcontent.com/pod-product-compliance
Lightning Source LLC
Chambersburg PA
CBHW071328190426
43193CB00041B/1002